Random Vibration and Spectrum Analysis

Jimin Zhang · Hechao Zhou

Random Vibration and Spectrum Analysis

CHINA RAILWAY PUBLISHING HOUSE CO., LTD.

Springer

Jimin Zhang
Institute of Rail Transit
Tongji University
Shanghai, China

Hechao Zhou
Tongji University
Shanghai, China

ISBN 978-981-95-0190-8 ISBN 978-981-95-0191-5 (eBook)
https://doi.org/10.1007/978-981-95-0191-5

Jointly published with China Railway Publishing House Co., Ltd.
The print edition is not for sale in China Mainland. Customers from China Mainland please order the print book from: China Railway Publishing House Co., Ltd.
ISBN of the Co-Publisher's edition: 978-711-33-1434-7

This Springer imprint is published by the registered company Springer Nature Singapore Pte Ltd.
The registered company address is: 152 Beach Road, #21-01/04 Gateway East, Singapore 189721, Singapore

Preface

For nearly two decades, the editor Prof. Zhang Jimin has been immersed in the instruction of graduate-level courses on random vibration and spectrum analysis, tirelessly seeking a more fitting textbook to serve as the primary educational resource. Despite this pursuit, it has become apparent that the content of current reference materials varies significantly, with many of them having been published long ago. Regrettably, there exists no direct reference book on random vibration and spectrum analysis that is up-to-date, either domestically or internationally. Notable works in the field include Zhu Weigiu's *Random Vibration*, published in 1992, Gan Youchen's *Basic Theory and Applications of Random Vibration*, published in 1982, and Christian Lalanne's *Random Vibration* (Volume III), published in 2002.

On the basis of theory, this book is a tribute to harness the unique characteristics of random vibrations and cross-signal analysis techniques, drawing from the core principles of various disciplines such as theoretical mechanics, materials science, mechanical vibrations, statistical analysis, and engineering mathematics. This book is crafted to explore and tackle the intricacies of random vibration response analysis within mechanical systems, with a particular focus on railway vehicles and other engineering domains.

In terms of structure, this book is thoughtfully designed, encompassing the essentials of vibration theory, probability theory, random variables, stochastic processes, and spectrum analysis. It examines the transmission characteristics of random vibrations in both single-degree and multi-degree freedom systems, with a specific focus on the random vibrations encountered in railway vehicles. The book elucidates key theoretical frameworks and methodologies, including integral transformations, correlation analysis for random vibrations, spectrum analysis techniques, the dynamics of random excitation-response in linear systems, the computation of statistical properties of random responses, and a holistic analysis of random vibrations.

The main content and distinctive features of this book are as follows:

(1) A careful selection of fundamental theoretical knowledge of vibration, presented in an integrated manner within the book, to meet the educational and research needs of readers from diverse backgrounds.

(2) A deliberate selection of content on random vibration and spectrum analysis, seamlessly integrated rather than merely aggregated or haphazardly collected.
(3) A balance of universality and professionalism, addressing the specifics of random vibration problems in the professional field while also considering general applicability.

The aim of this book is to organically and concisely incorporate the basic theoretical knowledge of vibration into the textbook on random vibration and spectrum analysis; to achieve a harmonious blend of generality and professionalism; and to design teaching cases that are rooted in professional practice.

This book is written by Prof. Zhang Jimin and Dr. Zhou Hechao from Tongji University. Prof. Zhang Jimin is responsible for Chapters 1 through 5, and Dr. Zhou Hechao is responsible for Chapters 6 and 7. The publishing of this book is proudly sponsored by the Graduate Book Construction Project of Tongji University. We would like to extend our heartfelt gratitude to Tongji University and the Institute of Railway and Urban Rail Transit at Tongji University for their funding.

Shanghai, China
April 2024

Prof. Jimin Zhang

Contents

Chapter 1
Introduction

1.1 Basic Concepts

Deterministic vibration, also known as the vibration rule, refers to the type of vibration that can be precisely described by a definite mathematical relationship. For instance, consider a body suspended by a linear spring on a fixed board, as depicted in Fig. 1.1.

The law of motion for such a system is completely determined, making deterministic vibration the primary subject of study in classical vibration theory.

On the other hand, random vibration employs advanced mathematical tools and random process theory to address vibration problems. From a mechanical perspective, it represents an evolution of classical vibration theory. Mathematically, it is a significant application of random process theory within the field of vibration. Practically, it primarily concerns engineering problems.

Determined and random vibrations are distinct. For example, the unevenness of a road or rail produces irregular impacts of varying intervals, magnitudes, and frequencies on the wheels, thereby causing the vehicle to vibrate randomly. The movement of a vehicle at a single point is illustrated in Fig. 1.2.

The vibration varies over time and cannot be described by a deterministic function. It is unpredictable, meaning that the vibration at any given moment is uncertain. Comparing two recorded signals (or images) taken at equal time intervals throughout the vibration process, one will find that they are never exactly the same; that is, the record is non- repeating. The first characteristic of this type of vibration is its non-repeatability and unpredictability, in other words, the vibration is random.

This does not imply that the vibration lacks order; rather, it cannot be described in terms of deterministic functions. For instance, if road conditions remain constant, certain characteristics of the vibration can be expected to remain unchanged over a certain period of time. If the recording time Δt is sufficient, the average value of the signal, both inside and outside (as denoted by AB in Fig. 1.2), will be the same. And

J. Zhang and H. Zhou, *Random Vibration and Spectrum Analysis*,
https://doi.org/10.1007/978-981-95-0191-5_1

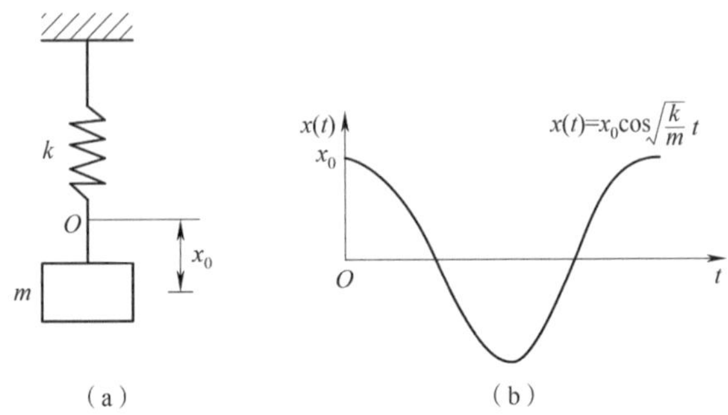

Fig. 1.1 Mass suspended by a spring

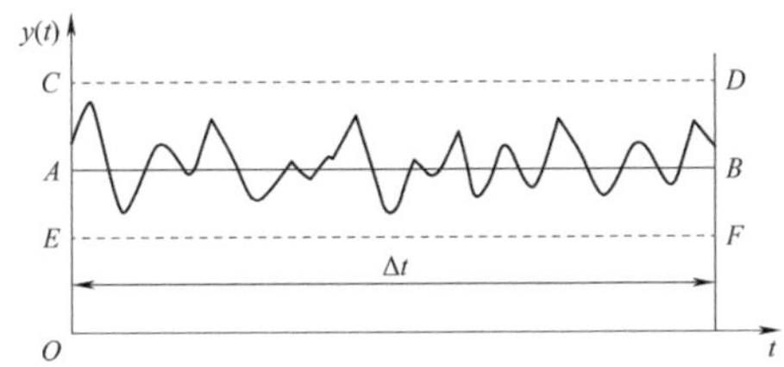

Fig. 1.2 A random vibration time history

the peaks of the signal $y(t)$ will still fall within the range of CD and EF, except in very unusual circumstances.

Therefore, by employing statistical concepts, a more accurate description of the basic characteristics can be made based on the recorded data. This indicates that the vibration exhibits a certain statistical regularity and can be described by stochastic process theory, which is another defining characteristic of random vibration.

According to the characteristics of random vibration, the behavior of actual systems can be investigated to determine whether they exhibit regular vibration patterns. However, it is important not to equate the complexity of a wave with randomness. In other words, random vibration is governed by "probability", not by complication.

1.2 Several Concepts of Random Vibration

1. **Random Variable**

A random variable is a quantity whose instantaneous value cannot be predicted with certainty. Knowledge of the variable's values before time t does not enable one to deduce its value at time t. An example of this is the Brownian movement of a particle.

2. **Random Process**

Consider, for example, the acceleration recorded at a specific point on a train's body as it travels on a good rail between two cities. For a single journey, the recorded acceleration conforms to the definition of a random variable. The vibration characterized by this acceleration is referred to as random or stochastic. If n journeys are undertaken, one obtains as many different $a_i(t)(i = 1, 2, \ldots, n)$ curves, each having a random nature. We define a random process or stochastic process as the ensemble of the time functions $\{a_i(t)\ (i = 1, 2, \ldots, n)\}$ for t ranging from $-\infty$ to $+\infty$. This ensemble is defined by its statistical properties.

3. **Random Vibration**

Random movements are not erratic in the common sense of the term, follow a well-defined law. They possess specific properties and can be described by a probability law. The principal characteristic of random vibration is that it simultaneously excites all frequencies of an object. Unlike deterministic vibrations, random vibrations consist of a continuous range of frequencies, with the signal's amplitude and phase varying with time in a random manner. Random functions are sometimes defined as a continuous distribution of sinusoids of all frequencies, with amplitudes and phases that vary randomly with time.

1.3 Random Vibration in the Vehicle Field

When examining vehicle vibrations, two types of excitations are typically considered: one is deterministic, such as the periodicity of joints in railway vehicles, which can be used as a deterministic excitation when random factors are ignored. The response of the track vehicle can be obtained through a simulation model of the track vehicle entering and exiting the turnout. The other type is randomness, such as line or road unevenness, which requires processing the vehicle's response under the random excitation of the line or road in both the time and frequency domains. In an engineering context, when dealing with the random vibration of vehicles, the following assumptions are generally made:

(1) The external excitation and response of a vehicle can usually be considered as a stationary random process. In essence, during multiple random test detections, the probability statistical law of excitation and response at a certain time remains the same as the probability statistical law of delayed time, and it does not change with time shift.
(2) Furthermore, there is the assumption of ergodicity for each state, meaning that the probability statistical law of external excitation and response received by a vehicle in a random test can reflect the statistical law obtained from multiple test sets. This allows researchers to determine the random vibration statistical characteristics from a sufficiently long sample of a single vehicle's time history records.
(3) Random vibration arises from various factors, and the probability density of certain parameters in random vibration typically conforms to the assumption of a normal distribution (Gaussian distribution). The assumption of a normal distribution simplifies and facilitates statistical analysis for researchers.

1.4 Random Vibration Knowledge System

The knowledge system of random vibration is founded on the dynamics theory of classical Newtonian mechanics and the deterministic dynamics theory of deterministic vibration. It evolves by integrating stochastic processes with kinematic equations dominated by second-order differential equations to derive random vibration, and further integrates stochastic processes with stochastic dynamics differential equations to develop into the theory of random dynamics. The specific framework system is illustrated in Fig. 1.3.

Random vibration and spectrum analysis primarily encompass three areas of knowledge: random vibration, random processes, signal analysis and processing, as depicted in Fig. 1.4.

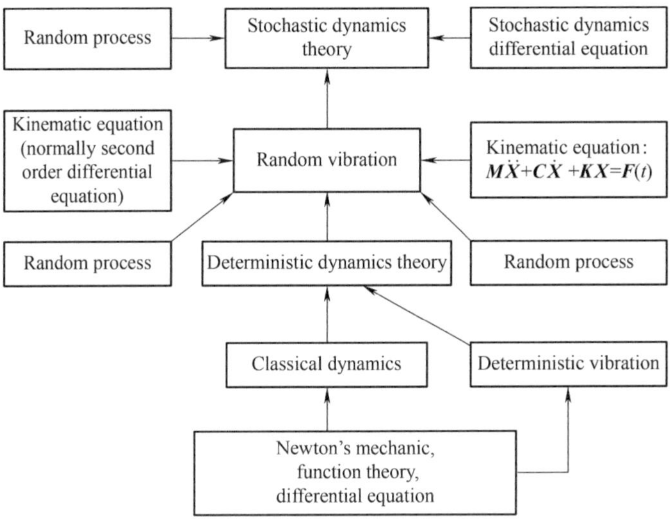

Fig. 1.3 The specific framework system of the random vibration

Fig. 1.4 Three areas of knowledge of random vibration and spectrum analysis

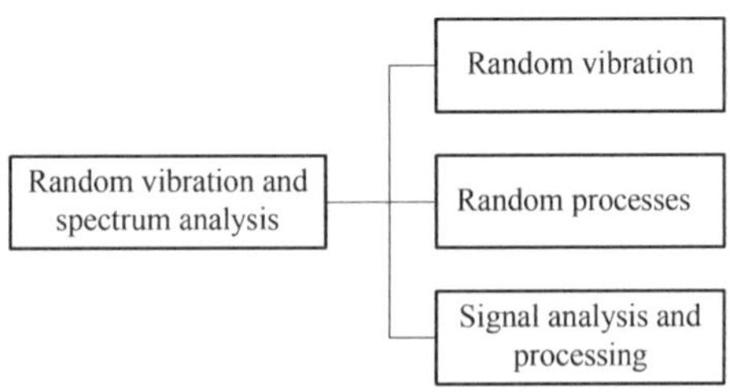

1.5 Research Methods of Random Vibration

Due to the randomness and the frequency content of vibrations, the random vibration can only be studied using statistical methods applied to signals over time or by using frequency domain plots (spectra).

There are four principal approaches to analyse random vibration:

(1) Analysis of the ensemble statistical properties of the process.
(2) Methods of correlation.
(3) Spectrum analysis.
(4) Analysis of the statistical properties of the signal with respect to time.

The analysis process for random vibration is given in Fig. 1.5.

The main parameters most frequently used in practice include:

(1) The root mean square (RMS) value of the signal and, if applicable, its variation as a function of time.
(2) The distribution of the signal with respect to time.
(3) The power spectral density.

Exercise 1

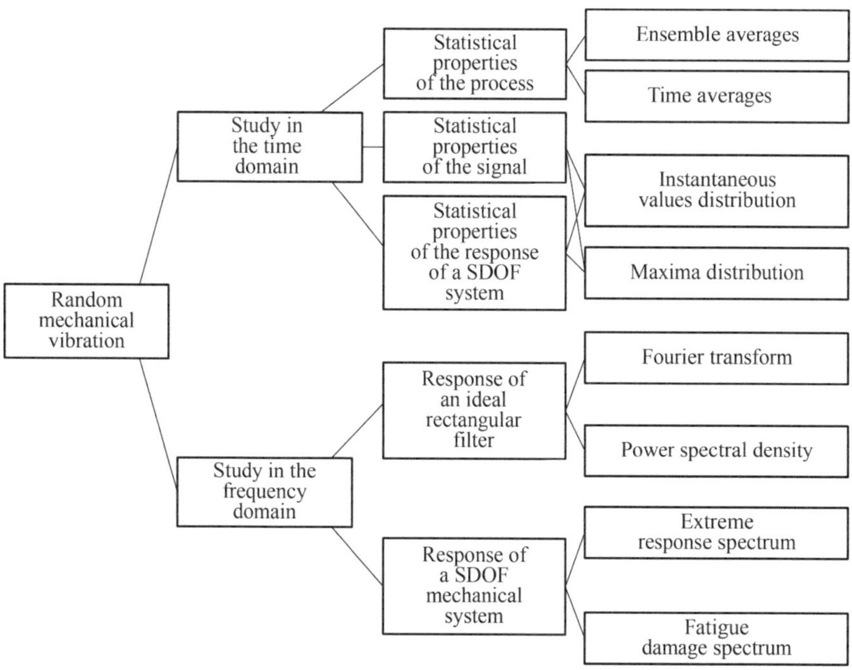

Fig. 1.5 The analysis process for random vibration

E1.1: Could you elucidate the fundamental concepts of vibration and random vibration, highlighting their similarities and contrasting characteristics?

E1.2: Please provide examples of random vibrations as they manifest in trains and rail lines within the rail transit field.

E1.3: What are the three essential knowledge systems that underpin the study of random vibration and spectrum analysis?

E1.4: Could you briefly outline the four principal approaches utilized in the analysis of random vibration?

Chapter 2
Vibration

Human beings reside in a world teeming with vibrations. Vibration plays a pivotal role in various fields such as mechanics, acoustics, electrical science, bioengineering, and automatic control.

This chapter presents an introductory overview of vibrations. It introduces the fundamental concepts of degrees of freedom and the distinctions between discrete and continuous systems. Additionally, it describes the elementary components of vibrating systems. The chapter outlines the various classifications of vibrations, including free and forced, undamped and damped, linear and nonlinear, deterministic and random vibrations. It also delineates the steps involved in the vibration analysis of engineering systems and introduces essential definitions and concepts of vibration.

The chapter delves into the concept of harmonic motion, illustrating its representation through vectors and complex numbers. It provides foundational definitions and terminology related to harmonic motion, such as cycle, amplitude, period, frequency, phase angle, and natural frequency. The vibration analysis procedure is introduced with illustrative examples.

2.1 Description of Vibration

2.1.1 General Concept of Vibration

1. **General Definition**

Vibration refers to the repetitive motion of an object along the same path or position. In essence, any motion that exhibits periodicity is termed vibration or oscillation. For example, a pendulum's swing and a plucked string's vibration are quintessential examples of this phenomenon.

J. Zhang and H. Zhou, *Random Vibration and Spectrum Analysis*,
https://doi.org/10.1007/978-981-95-0191-5_2

2. **Key Points**

Periodic vibration is characterized by its balance point, repeatability, and periodicity, with a defined period *T*, allowing the motion state to be entirely replicated.

Non-periodic vibration, on the other hand, lacks the ability to duplicate its motion state completely within a period *T*.

3. **Elementary Components of Vibration System**

A typical vibration system comprises components that store potential energy (such as springs or elastic elements), kinetic energy (masses or inertial elements), and components that dissipate energy (dampers). These systems usually alternate the transfer of energy between potential and kinetic forms. In damped systems, some energy is lost, and an external source is required to maintain a steady state of vibration.

4. **Classification of Mechanical Vibration**

Vibrations can be categorized in multiple ways, with the following being the most significant classifications:

(1) Free and forced vibration: free vibration occurs when a system vibrates on its own after an initial disturbance, without any subsequent external forces. An example is the oscillation of a simple pendulum. Forced vibration occurs when an external force acts upon the system, such as the vibrations in a diesel engine. Resonance occurs when the external force's frequency matches one of the system's natural frequencies.
(2) Undamped and damped vibration: undamped vibration is characterized by no energy loss during oscillation, whereas damped vibration involves energy dissipation.
(3) Linear and nonlinear vibration: linear vibration is defined by the linear behavior of the system's components, such as springs and dampers. Nonlinear vibration arises when any component exhibits nonlinear characteristics. The governing equations for these systems are linear or nonlinear, respectively.
(4) Deterministic and random vibration: deterministic vibration is characterized by known excitation values or magnitudes, while random vibration involves unpredictable excitation values that may exhibit statistical regularities, such as mean and mean square values.

Additionally, vibrations can be classified based on their causes (free, forced, self-excited, parametric), degrees of freedom (single, multi), and displacement types (angular, linear).

Most vibrations can be decomposed into harmonic vibrations, which are fundamental and ubiquitous in physical phenomena.

2.1.2 Harmonic Vibration

1. **Representation of Harmonic Motion and Its Model**

Oscillatory motion can be either regular and periodic or irregular. Periodic motion, which repeats at equal time intervals, is referred to as harmonic motion when it follows a simple pattern.

Harmonic vibration is characterized by an object's displacement from its equilibrium position varying according to cosine (or sine) laws over time.

Let's examine the characteristics and representation of a harmonic vibration, as depicted in Fig. 2.1, which shows a spring-ball system on a frictionless plane.

Restoring force is the force acting on a particle in harmonic motion, which is proportional to the displacement and directed oppositely.

When the mass of the spring is negligible and there is no deformation of the object, the dynamic characteristics of harmonic vibration, derived from Newton's second law, are as follows:

$$F = -kx \tag{2.1}$$

Assuming: $k/m = \omega^2$

$$a = \frac{F}{m} = -\frac{k}{m}x$$

$$\Rightarrow a = \frac{d^2x}{dt^2} = -\omega^2 x, \quad \frac{d^2x}{dt^2} + \omega^2 x = 0 \quad \text{(Kinematic characteristics)}$$

The solution for displacement is given by

$$x = A\cos(\omega t + \varphi_0) \tag{2.2}$$

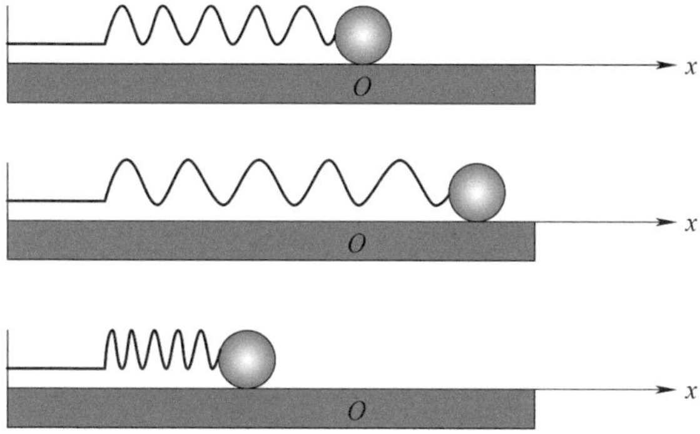

Fig. 2.1 A harmonic vibration example

$$\frac{d^2x}{dt^2} + \omega^2 x = 0 \quad \text{or} \quad x = Ae^{i(\omega t + \varphi_0)} \tag{2.3}$$

The kinematic characteristics of harmonic vibration dictate that an object's acceleration is proportional to its displacement, with the direction being opposite, and the displacement changes according to the cosine law.

$$\text{Velocity}: \quad v = \frac{dx}{dt} = -\omega A \sin(\omega t + \varphi_0)$$

$$\text{Acceleration}: \quad a = \frac{d^2x}{dt^2} = -\omega^2 A \cos(\omega t + \varphi_0)$$

Example 2.1 Consider a flat-bottomed boat with mass m, draft depth h, water density ρ, gravity accelation g and average area S (as shown in Fig. 2.2). Neglecting water resistance, determines the vibration period of the vessel in the vertical direction.

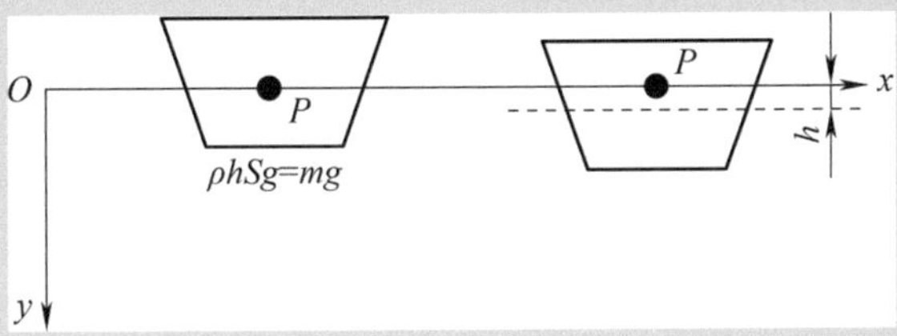

Fig. 2.2 A boat floats in the water

Solution

When the boat is stationary, the buoyancy force balances the gravitational force.

The vessel, at any location, takes the water surface as the coordinate origin, with the vertical axis as the y-axis and the vessel's displacement as y.

The combined force acting on the vessel when its displacement y is as follows.

$$F = -(h + y)\rho S g + mg = -y\rho S g$$

The vessel undergoes harmonic vibration in the vertical direction, with an angular frequency $\omega = \sqrt{\frac{\rho S g}{m}}$ and period $T = \frac{2\pi}{\omega} = 2\pi\sqrt{\frac{m}{\rho g S}}$.

Because $m = \rho S h$, we get $T = 2\pi\sqrt{h/g}$.

The relation of displacement, velocity, and acceleration in harmonic vibration is illustrated in Fig. 2.3.

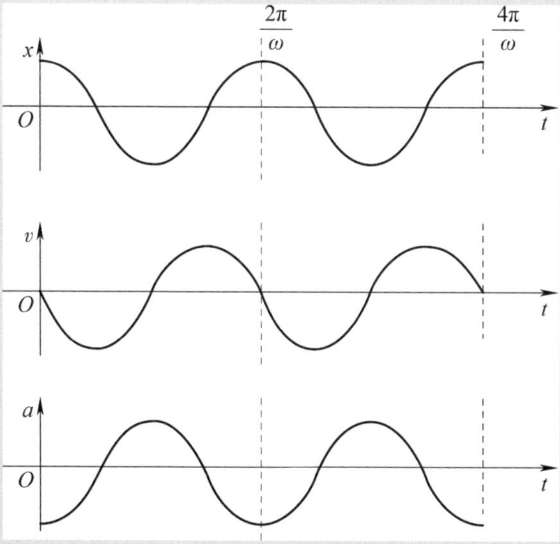

Fig. 2.3 The relation of displacement, velocity, acceleration in harmonic vibration

There is a problem: how to calculate the constants A and φ_0?
From the initial conditions, when $t = 0,\quad x = x_0,\quad v = v_0$, we get

$$x_0 = A\cos\ \varphi_0,\quad v_0 = -\omega A\sin\ \varphi_0$$

$$\Rightarrow A = \sqrt{x_0^2 + (v_0/\omega)^2},\quad \varphi_0 = \arctan\left(-\frac{v_0}{\omega x_0}\right)$$

From $-\pi$ to $+\pi$, there are typically two values. It can be determined by

$$v_0 = -\omega A\sin\ \varphi_0$$

Example 2.2 Solve for the magnitude, period, frequency, and phase of the harmonic motion.

Solution

(1) Magnitude: the magnitude is the absolute value of the maximum displacement of the object from its equilibrium position.

Initial condition is

$$A = \sqrt{x_0^2 + (v_0/\omega)^2}$$

(2) Period and frequency: the period is the time taken for an object to complete a full cycle of motion.

$$x = A\cos(\omega t + \varphi_0) = A\cos[\omega(T + t) + \varphi_0]$$

$$\Rightarrow T = 2\pi/\omega$$

Frequency: the number of complete cycles the object undergoes in a unit of time.

$$\gamma = 1/T = \omega/2\pi$$

Angular frequency: the number of complete cycles the object undergoes in seconds.

$$\omega = 2\pi/T = 2\pi\gamma$$

For a spring oscillator, because $\omega = \sqrt{k/m}$, we get

$$T = 2\pi\sqrt{m/k}, \quad \gamma = \frac{1}{2\pi}\sqrt{k/m}$$

Using the above relationships, the harmonic vibration can be represented as

$$x = A\cos(2\pi t/T + \varphi_0), \quad or \quad x = A\cos(2\pi\gamma t + \varphi_0)$$

(3) Phase and initial phase: phase $\omega t + \varphi_0$ defines the state of the harmonic motion.The initial phase φ_0 is the phase at $t = 0$. The phase concept can be used to compare the differences between two harmonic vibrations in their vibration steps.

There are two harmonic vibrations with the same frequency, presented as

$$x_1 = A_1 \cos(\omega t + \varphi_{10})$$
$$x_2 = A_2 \cos(\omega t + \varphi_{20})$$

The difference between the above two is

$$\Delta\varphi = (\omega t + \varphi_{20}) - (\omega t + \varphi_{10}) = \varphi_{20} - \varphi_{10} \tag{2.4}$$

Discussion:

(a) When $\Delta\varphi = 2k\boldsymbol{\pi}$**,** it is called the same phase.
(b) When $\Delta\varphi = (2k + 1)\boldsymbol{\pi}$**,** it is called the reverse phase.
(c) When $\Delta\varphi > 0$, the second vibration is ahead of the first phase.
(d) When $\Delta\varphi < 0$, the second vibration is behind of the first phase.

The phase can be used to study the changes in physical quantities such as displacement, velocity, and acceleration of harmonic vibration. For example:

$$x = A\cos(\omega t + \varphi_0)$$
$$v = -v_m \sin(\omega t + \varphi_0) = v_{\mathrm{m}} \cos(\omega t + \varphi_0 + \pi/2)$$
$$a = -a_m \cos(\omega t + \varphi_0) = a_m \cos(\omega t + \varphi_0 + \pi)$$

The phase of velocity is ahead of the phase of displacement. The phase of acceleration is 90°ahead of the phase of displacement.

Example 2.3 An object oscillates along the x-axis by harmonic vibration. The magnitude $A = 0.12$ m, period $T = 2$ s. When $t = 0$, the displacement $x =$ 0.06 m, and the direction is along the positive x-axis.

Find:

(1) Harmonic vibration presentation;
(2) When $t = T/4$, the displacement, velocity, and acceleration;
(3) The first time the object passes the equilibrium position when moving from $x = -\ 0.06$ m to the negative direction of the x-axis.

Solution

(1) Taking the equilibrium point as the coordinate origin, the harmonic vibration equation is

$$x = A\cos(\omega t + \varphi_0)$$

Initial conditions are $t = 0$, $x_0 = 0.06$ m, we can get

$$0.12\cos\ \varphi_0 = 0.06 \Rightarrow \varphi_0 = \pm\pi/3$$

From the initial condition $v_0 = -\omega A\ \sin\varphi_0 > 0$, we can get

$$\varphi_0 = -\pi/3$$
$$\Rightarrow x = 0.12\cos(\pi t - \pi/3)$$

(2) From (1)

$$v = \frac{dx}{dt} = -0.12\pi\ \sin(\pi t - \pi/3)$$
$$a = \frac{dv}{dt} = -0.12\pi^2\cos(\pi t - \pi/3)$$

At $t = T/4 = 0.5$ s,

$$x = 0.12\cos(\pi \times 0.5 - \pi/3) = 0.104\,\text{m}$$
$$v = -0.12 \times \pi\ \sin(\pi \times 0.5 - \pi/3) = -0.18\,\text{m}\,\text{s}^{-1}$$
$$a = -0.12 \times \pi^2\cos(\pi \times 0.5 - \pi/3) = -1.03\,\text{m}\,\text{s}^{-2}$$

(3) When $x = -0.06$ m, suppose the time is t_1, get

$$\cos(\pi t_1 - \pi/3) = -1/2$$
$$\Rightarrow \pi t_1 - \pi/3 = 2\pi/3, 4\pi/3$$

Because the velocity is negative at this time, $4\pi/3$ is given up, we get $t_1 = 1$ s.

Assuming the time is t_2, when the object returns to the equilibrium point and its phase is $3\pi/2$.

Therefore, the first time passing the equilibrium point of $x = -0.06$ m is

$$\Delta t = t_2 - t_1 = 0.83\ \text{s}$$

Another solution: the related phase difference from t_1 to t_2 is

$$\Delta\varphi = 3\pi/2 - 2\pi/3 = 5\pi/6 \quad \Rightarrow \Delta t = \Delta\varphi/\omega = 0.83\ \text{s}$$

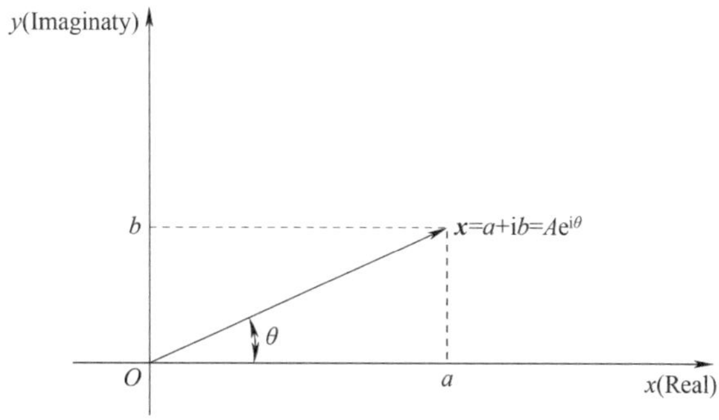

Fig. 2.4 Representation of harmonic vibration in the complex plane

2. **Complex Number Representation of Harmonic Motion**

Harmonic motion is conveniently represented using complex numbers, see Fig. 2.4. Any vector in the xy-plane can be expressed as a complex number:

$$\boldsymbol{x} = a + \mathrm{i}b \tag{2.5}$$

The complex number $\boldsymbol{x}$ can be rewritten as:

$$\boldsymbol{x} = A\cos\theta + \mathrm{i}A\sin\theta \tag{2.6}$$

$$A = \sqrt{a^2 + b^2}, \quad \theta = \arctan(b/a)$$

where A is the modulus or absolute value of the vector, and θ is the argument or the angle between the vector and the x-axis.

2.2 Vector Diagram Method of Harmonic Vibration

Utilizing the complex number representation, the vector $\boldsymbol{x}$ can be expressed as:

$$\boldsymbol{x} = A\mathrm{e}^{\mathrm{i}\omega t} \tag{2.7}$$

Rotation vector: a vector of length equal to the amplitude A is counter clockwise rotated around the O point in the plane, with an angular velocity equal to the angular frequency of the harmonic vibration. The differentiation of the harmonic vibration given by Eq. (2.8) with respect to time yields:

$$\frac{\mathrm{d}\boldsymbol{x}}{\mathrm{d}t} = \frac{\mathrm{d}}{\mathrm{d}t}\left(A\mathrm{e}^{\mathrm{i}\omega t}\right) = \mathrm{i}\omega A\mathrm{e}^{\mathrm{i}\omega t} = \mathrm{i}\omega\boldsymbol{x} \tag{2.8}$$

$$\frac{\mathrm{d}^2\boldsymbol{x}}{\mathrm{d}t^2} = \frac{\mathrm{d}}{\mathrm{d}t}\left(\mathrm{i}\omega A\mathrm{e}^{\mathrm{i}\omega t}\right) = -\omega^2 A\mathrm{e}^{\mathrm{i}\omega t} = -\omega^2\boldsymbol{x} \tag{2.9}$$

The displacement, velocity, and acceleration can be expressed as:

$$\text{Displacement}:\quad x = Re(A\mathrm{e}^{\mathrm{i}\omega t}) = A\cos(\omega t) \tag{2.10}$$

$$\text{Velocity}:\quad v = Re(\mathrm{i}\omega A\mathrm{e}^{\mathrm{i}\omega t}) = -\omega A\sin(\omega t) = \omega A\cos\left(\omega t + \frac{\pi}{2}\right) \tag{2.11}$$

$$\text{Acceleration}:\quad a = Re\left(-\omega^2 A\mathrm{e}^{\mathrm{i}\omega t}\right) = -\omega^2 A\cos(\omega t) = \omega^2 A\cos(\omega t + \pi) \tag{2.12}$$

Using the rotation vector method, we can intuitively understand the physical significance of the harmonic vibration expressions. Figure 2.5 shows two harmonic vibrations of the same frequency composed one vibration, with the vector represented by $\boldsymbol{A}$. Figure 2.6 illustrates the elements of the vector $\boldsymbol{A}$, including the magnitude of A, the angular velocity of ω and the counter clockwise rotation direction of the vector.

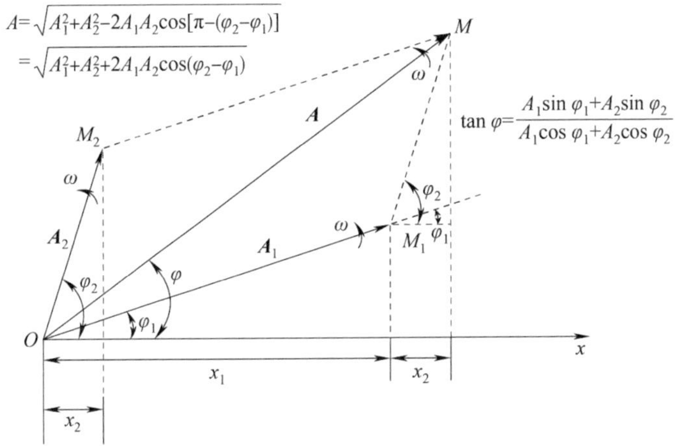

Fig. 2.5 Two harmonic vibrations of the same frequency composed one vibration

Fig. 2.6 The elements of the vector $\boldsymbol{A}$

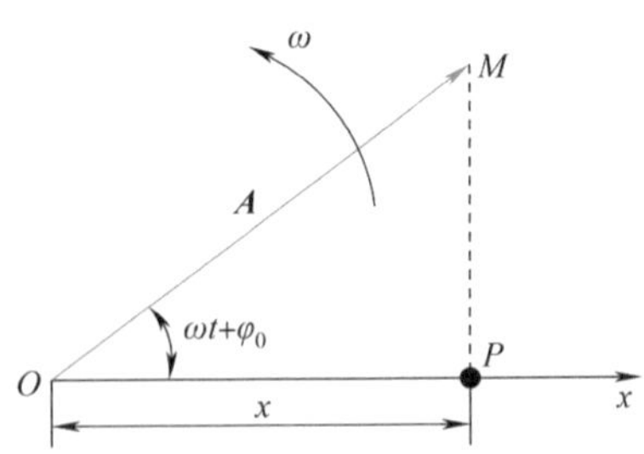

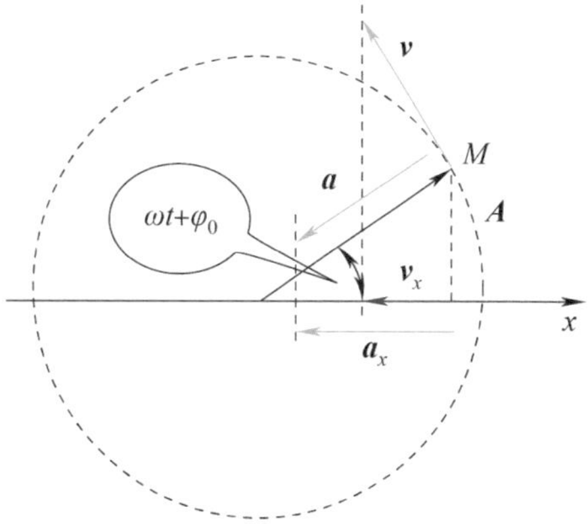

Fig. 2.7 The unit circle representation of the vector

The angle between the reference direction x of r and the vector $\boldsymbol{A} = \omega t + \varphi_0$. The movement of point P, projected on the x-axis(P) of point M, is shown in

$$x = A\cos(\omega t + \varphi_0)$$

$\boldsymbol{v}_x$, $\boldsymbol{a}_x$ are the projected on the x-axis of velocity and acceleration vectors from Fig. 2.7.

For point M:

$$v_m = \omega A$$
$$a_m = \omega^2 A$$

Two harmonic vibrations of the same frequency:

$$x_1 = A_1\cos(\omega t + \varphi_1), \quad x_2 = A_2\cos(\omega t + \varphi_2) \tag{2.13}$$

The phase difference is

$$\Delta\varphi = (\omega t + \varphi_2) - (\omega t + \varphi_1) = \varphi_2 - \varphi_1 \tag{2.14}$$

Using the rotation vector method, the expression is shown in Fig. 2.8.

2.3 Several Common Harmonic Vibrations

Oscillatory motion can be regular, as in the case of a simple pendulum, or irregular, as during an earthquake's ground motion. If the motion repeats at equal time intervals, it is termed periodic motion. The simplest form of periodic motion is harmonic motion. The following are examples of harmonic vibrations.

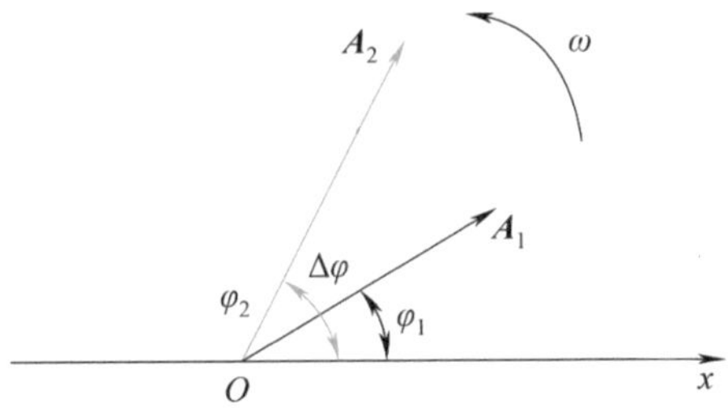

Fig. 2.8 Two harmonic vibrations of the same frequency

2.3.1 Pendulum

The simple pendulum is depicted in Fig. 2.9. The ball's mass is m, and the pendulum's length is l.

The object is subject to a composite external torque, as shown in Eqs. (2.15) and (2.16).

$$M = -mgl\sin\theta \tag{2.15}$$

$$\sin\theta = \theta - \frac{\theta^3}{3!} + \frac{\theta^5}{5!} - \cdots \tag{2.16}$$

When the amplitude is very small (less than the radius), we take

$$\sin\theta \approx \theta, \quad M = -mgl\theta \tag{2.17}$$

According to the rotational law, we get

$$\frac{d^2\theta}{dt^2} = \frac{M}{J} = -\frac{mgl\theta}{ml^2} = -\frac{g\theta}{l} \tag{2.18}$$

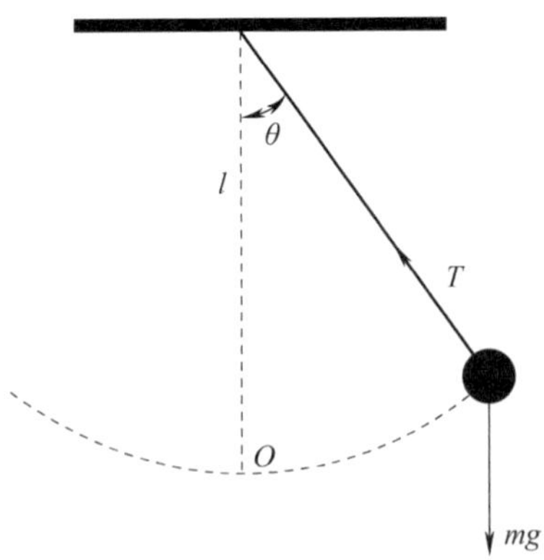

Fig. 2.9 The simple pendulum

Given $\omega^2 = g/l$, we obtain $T = 2\pi/\omega = 2\pi\sqrt{g/l}$.
The presentation can be written as

$$\theta = \theta_m \cos(\omega t + \varphi_0)$$

From the initial condition, we get θ_m and φ_0.
The relationship between the period and the angular magnitude is

$$T = T_0\left(1 + \frac{1}{2^2}\sin^2\frac{\theta_m}{2} + \frac{1}{2^2}\frac{3^2}{4^2}\sin^4\frac{\theta_m}{2} + \cdots\right) \tag{2.19}$$

The vibration of the pendulum is period, when the amplitude is very small.
According to the series formulas of the period, the period T can be calculated to any desired precision.

2.3.2 Duplicate Pendulum

Figure 2.10 shows a dynamic motion system where a rigid body oscillates slightly around a fixed horizontal axis under the influence of gravity, known as a duplicate pendulum.

The mass center is C. The inertia is J with respect to the axle of O. The distance between point O and point C is h.

From the rotational law, we have

$$J\frac{d^2\theta}{dt^2} = -mgh\sin\theta \tag{2.20}$$

If the amplitude is small, we get

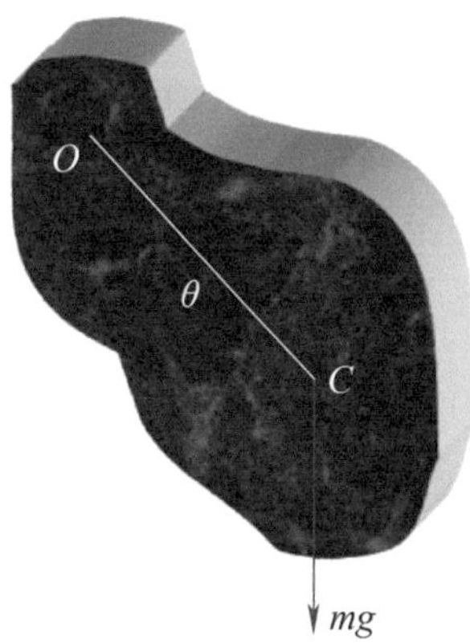

Fig. 2.10 The duplicate pendulum

$$J\frac{d^2\theta}{dt^2} = -mgh\theta \tag{2.21}$$

Given $\omega^2 = \frac{mgh}{J}$, we obtain

$$\frac{d^2\theta}{dt^2} + \omega^2\theta = 0$$

$$\Rightarrow T = \frac{2\pi}{\omega} = 2\pi\sqrt{\frac{J}{mgh}} \tag{2.22}$$

2.4 Energy of Harmonic Vibration

The Principle of Conservation of Energy states that a system is conservative if no energy is lost due to energy-consuming components. If no work is done on a conservative system by external forces, the system's total energy remains constant. Since a vibrating system's energy is partly potential and partly kinetic, the sum of these two energies remains constant. The following example of a simple harmonic oscillator illustrates this energy conservation.

The kinetic energy:

$$E_k = \frac{1}{2}mv^2 = \frac{1}{2}m\omega^2A^2\sin^2(\omega t + \varphi_0) \tag{2.23}$$

The potential energy:

$$E_p = \frac{1}{2}kx^2 = \frac{1}{2}kA^2\cos^2(\omega t + \varphi_0) \tag{2.24}$$

Total mechanical energy of the system:

$$E = E_k + E_p$$

The mechanical energy of the harmonic vibration is conserved, as shown by the equation $E = \frac{1}{2}kA^2$.

The average energy:

$$\overline{E}_k = \frac{1}{T}\int_0^T \frac{1}{2}m\omega^2A^2\sin^2(\omega t + \varphi_0)dt = \frac{1}{4}kA^2 \tag{2.25}$$

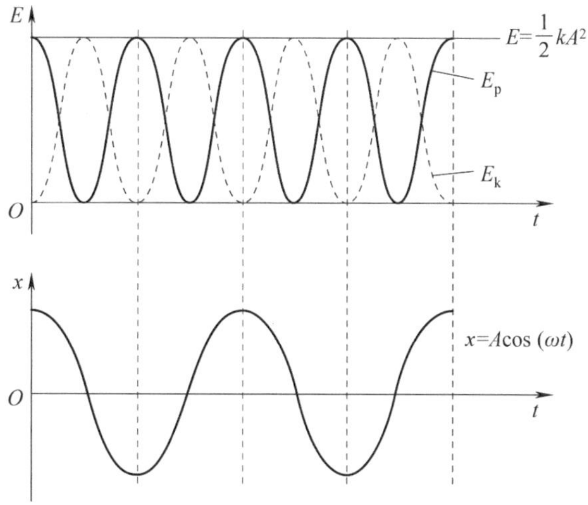

Fig. 2.11 The kinetic energy, potential energy and total energy to the harmonic oscillator

$$\overline{E}_p = \frac{1}{T}\int_0^T \frac{1}{2}kA^2\cos^2(\omega t + \varphi_0)dt = \frac{1}{4}kA^2 \tag{2.26}$$

$$\overline{E}_k = \overline{E}_p = E/2$$

The above results are established for any resonance system.

The changing curve of the kinetic energy, potential energy, and total energy of the harmonic oscillator follows (Fig. 2.11).

2.5 Damped Vibration

Undamped free vibration occurs when a vibrating object is unaffected by any damping force, and the vibration is solely under the influence of the restoring force.

Damped vibration occurs when the vibration is influenced by both the restoring force and the damping force.

Damping is the process of consuming the vibration system's energy, which include friction damping and viscoelastic damping.

The viscous damping force is proportional to the velocity or can be expressed as

$$F_c = -cv = c\frac{dx}{dt} \tag{2.27}$$

Spring oscillator under damping force including: elastic restoring force $-kx$, and resistance force F_c.

Kinematic equation:

$$m\frac{d^2x}{dt^2} = -kx - c\frac{dx}{dt}$$

Introduce the damping factor $\beta = c/(2m)$, natural frequency $\omega_0 = \sqrt{k/m}$, to get

$$\frac{d^2x}{dt^2} + 2\beta\frac{dx}{dt} + \omega_0^2 x = 0 \tag{2.28}$$

Under small damping condition ($\beta < \omega_0$) the solution to the differential Eq. (2.28) is

$$x = A_0 e^{-\beta t}\cos(\omega' t + \varphi_0') \tag{2.29}$$

where

$$\omega' = \sqrt{\omega_0^2 - \beta^2}$$

$$x = A_0 e^{-\beta t}\cos(\omega' t + \varphi_0')$$

where A_0 and φ_0' are integral constants, determined by the initial conditions. The cosine term in the Formula (2.29) reflects the periodic motion due to the elastic force and resistance force. The exponential term in the Formula (2.29) reflects the influence of damping on amplitude, as shown in Fig. 2.12.

Damping vibration is not periodic vibration and is not harmonic vibration, because the displacement is not time periodic function. However, the damping vibration still vibration still exhibits a regular cyclic motion.

The period of damped vibration is the time interval between two successive occurrences of reaching the maximum value.

$$T = \frac{2\pi}{\omega} = \frac{2\pi}{\sqrt{\omega_0^2 - \beta^2}} > \frac{2\pi}{\omega_0} \tag{2.30}$$

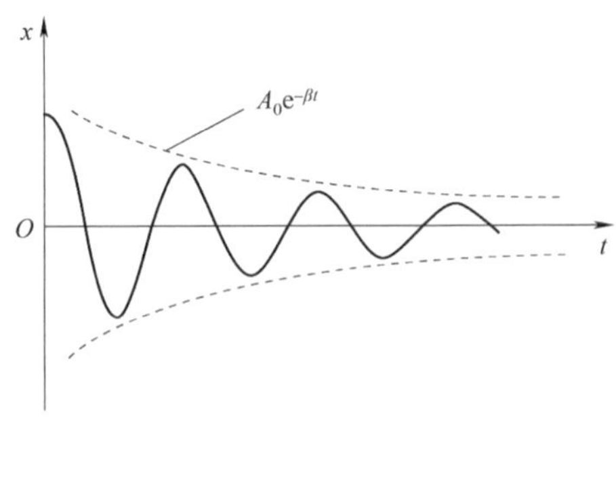

Fig. 2.12 The influence of damping on the amplitude

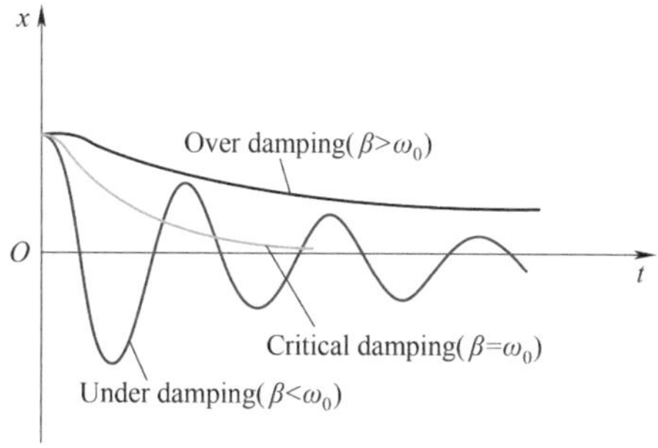

Fig. 2.13 The affection of different damping factors

It is evident that the vibration slows down due to damping. The magnitude of damped vibration is given by $A = A_0 e^{-\beta t}$.

The value of β determines the degree of attenuation of the damped vibration amplitude. Figure 2.13 shows the affection of different damping factors.

Three cases of damped vibration are over damping, under damping, and critical damping.

By controlling the damping value, different practicals needs can be met.

2.6 Forced Vibration and Resonance

2.6.1 Forced Vibration

When a system is subjected to an external force, often a periodic force, the resulting vibration is termed forced vibration. Resonance occurs when the frequency of the external force coincides with one of the system's natural frequencies, leading to dangerously large oscillations. Structural failures such as those in bogie frames, buildings, bridges. turbines, and airplane wings have been largely associated with resonance.

Driving force:

$$F = F_0 \cos(\omega t)$$

Kinematic equation with forced vibration:

$$m\frac{\mathrm{d}^2 x}{\mathrm{d}t^2} = -kx - c\frac{\mathrm{d}x}{\mathrm{d}t} + F_0 \cos(\omega t) \tag{2.31}$$

Assuming $\omega_0^2 = k/m$ and $\beta = c/(2m)$, we get

$$\frac{\mathrm{d}^2 x}{\mathrm{d}t^2} + 2\beta\frac{\mathrm{d}x}{\mathrm{d}t} + \omega_0^2 x = \frac{F_0}{m}\cos(\omega t) \tag{2.32}$$

For the small damping case, the solution can be written as

$$x = A_0 e^{-\beta t} \cos\left(\sqrt{\omega_0^2 - \beta^2} t + \phi_0'\right) + A\cos(\omega t + \phi_0) \tag{2.33}$$

After a period of time, the attenuation term $A_0 e^{-\beta t} \cos\left(\sqrt{\omega_0^2 - \beta^2 t} + \phi_0'\right)$ is neglected, and only the steady state term $A\cos(\omega t + \phi_0)$ is considered.

$$x = A\cos(\omega t + \varphi_0)$$
$$\Rightarrow A = \frac{F_0}{m\sqrt{\left(\omega_0^2 - \omega^2\right)^2 + 4\beta^2\omega^2}}, \quad \tan\varphi_0 = -\frac{2\beta\omega}{\omega_0^2 - \omega^2}$$

The velocity of the steady state is

$$v = \frac{dx}{dt} = v_m \cos(\omega t + \varphi_0 + \pi/2)$$
$$v_m = \frac{\omega F_0}{m\sqrt{\left(\omega_0^2 - \omega^2\right)^2 + 4\beta^2\omega^2}}$$

In forced vibration, the periodic driving force provides energy to the vibration system, while the system consumes energy due to damping. If they are equal, the system reaches a stable state of vibration.

2.6.2 Resonance

In forced vibration, when the amplitude of the external force is constant, the steady amplitude changes with the frequency of the driving force. When the angular frequency of the driving force equals a certain value, the displacement amplitude reaches its maximum, termed displacement resonance, as shown in Fig. 2.14.

From $\frac{dA}{d\omega} = 0$, get $\omega_\tau = \sqrt{\omega_0^2 - 2\beta^2}$.

Velocity resonance is the phenomenon of resonance in the forced vibration velocity under certain conditions.

From $\frac{dv_m}{d\omega} = 0$, get $\omega_r = \omega_0$.

Under the premise of very small damping, velocity resonance and displacement resonance can be considered the same.

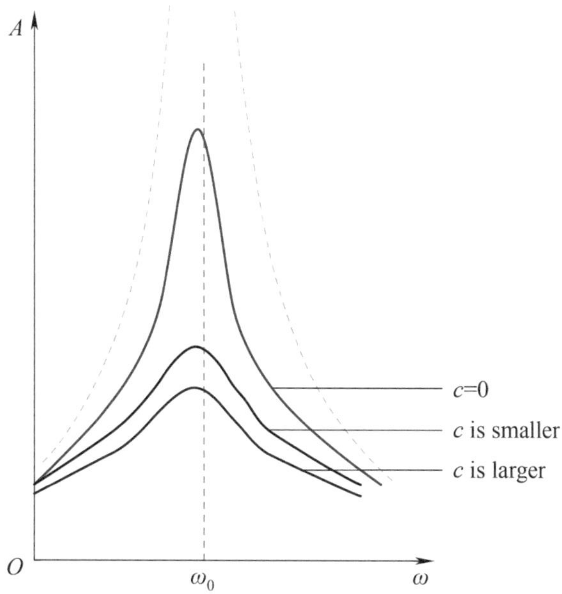

Fig. 2.14 Displcement amplitude under different damping factors

2.7 Synthesis of Two Harmonic Vibrations of the Same Direction

When a system undergoes multiple vibrations of the same frequency due to external periodic excitation, this vibration is sometimes referred to as multi-frequency resonance or resonance superposition. An example of the compound of harmonic vibrations is shown in Fig. 2.15.

1. **Same Frequency**

Suppose a particle participates in two harmonic vibrations with the same direction (e.g., along the x-axis) and independent frequencies. The two vibration displacements are:

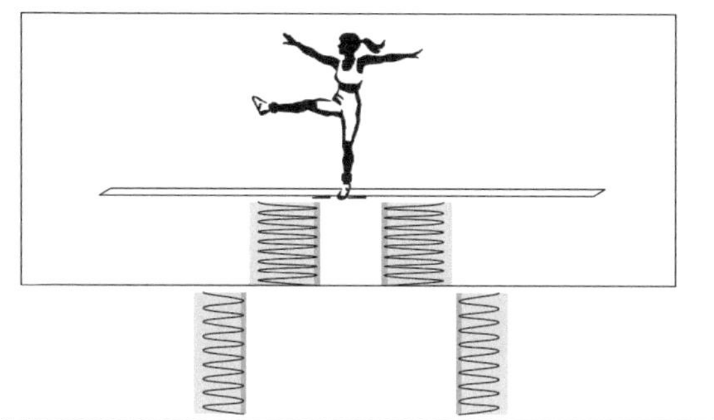

Fig. 2.15 Compound of harmonic vibrations

$$x_1 = A_1 \cos(\omega t + \varphi_{10}), \quad x_2 = A_2 \cos(\omega t + \varphi_{20}) \tag{2.34}$$

Compound displacement:

$$x = x_1 + x_2 = A \cos(\omega t + \varphi_0) \tag{2.35}$$

$$\Rightarrow A = \sqrt{A_1^2 + A_2^2 + 2A_1A_2 \cos(\varphi_{20} - \varphi_{10})}, \quad \tan\varphi = \frac{A_1 \sin\varphi_{10} + A_2 \sin\varphi_{20}}{A_1 \cos\varphi_{10} + A_2 \cos\varphi_{20}}$$

The synthesized vibration remains harmonic, with the same direction and frequency as the original vibrations. The synthesized vibration of two harmonic vibrations is illustrated in Fig. 2.5.

Discussion:

(1) When $\mathrm{d}f = f_{20} - f_{10} = 2kp$ [$k = 0$, positive and negative integers, $\cos(f_{20} - f_{10}) = 1$], we get

$$A = |\boldsymbol{A}_1 + \boldsymbol{A}_2|$$

When the vibrations are in the same phase, the amplitude reaches its maximum, as shown in Fig. 2.16.

(2) When $\mathrm{d}f = f_{20} - f_{10} = (2k + 1)\,p$ [$k = 0$, positive and negative integers, $\cos(f_{20} - f_{10}) = 0$], we get

$$A = |\boldsymbol{A}_1 - \boldsymbol{A}_2|$$

When the vibrations are out of phase, the compound amplitude is at its minimum, when $A_1 = A_2$ and $A = 0$, as shown in Fig. 2.17.

(3) In general, the compound amplitude lies between $|\boldsymbol{A}_1 + \boldsymbol{A}_2|$ and $|\boldsymbol{A}_1 - \boldsymbol{A}_2|$.

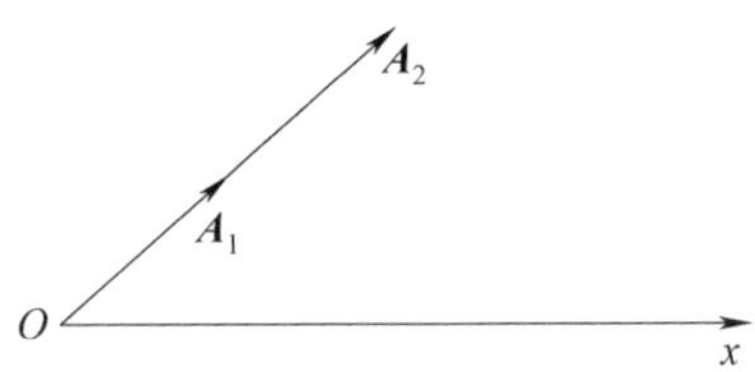

Fig. 2.16 The amplitude in the same phase

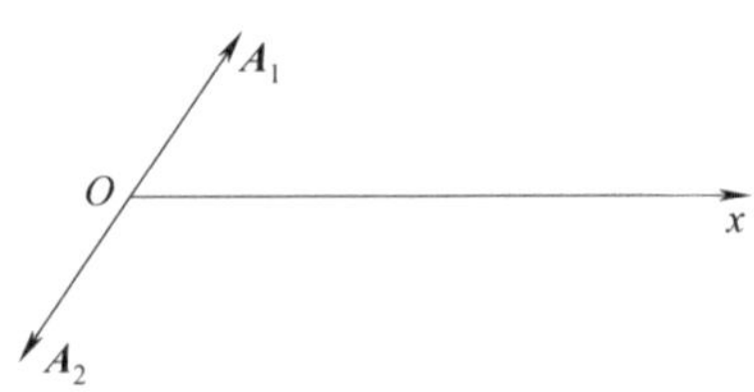

Fig. 2.17 The amplitude out of phase

Example 2.4 N number of harmonic vibrations in the same direction at the same frequency. Their amplitudes are equal, and the initial phases are $0, \alpha, 2\alpha, \ldots, (N-1)\alpha$, followed by a constant difference, the expression of vibration can be written as:

$$\begin{cases} x_1 = a\cos(\omega t) \\ x_2 = a\cos(\omega t + \alpha) \\ x_3 = a\cos(\omega t + 2\alpha) \\ \quad\vdots \\ x_N = a\cos[\omega t + (N-1)\alpha] \end{cases}$$

Solve for their amplitude and initial phase of the synthesized vibration.

Solution

Hints: the application of the rotation vector method can simplify the problem, avoiding cumbersome operations with trigonometric functions.

According to the vector synthesis method, the synthesis rotation vector diagram for N harmonic vibrations is shown in Fig. 2.18.

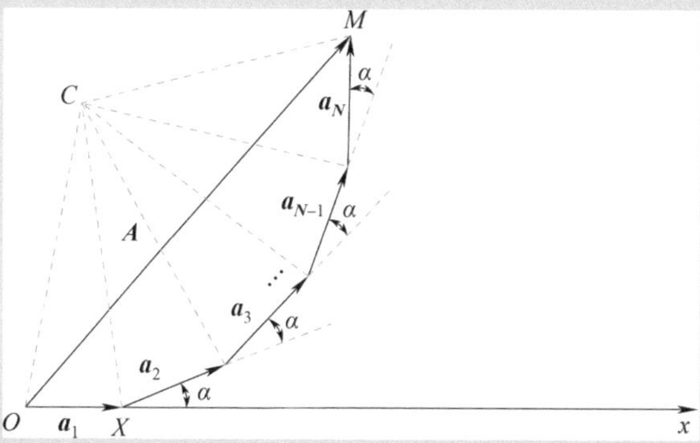

Fig. 2.18 *N* number harmonic vectors maglitude sum

For each vibration with the same amplitude and phase following a constant difference α, the beginning and end of the vector in the diagram form a circle centered at C. Using the geometric relationship, we get

$$\angle OCM = N\alpha$$

In triangle $\triangle OCM$, the length of OM is the displacement of the displacement vector, and the angle $\angle MOX$ is the initial phase of the vibration. Based on this, we obtain

$$A_i = 2\overline{OC}\sin\frac{N\alpha}{2}$$

Because $a = 2\overline{OC}\sin\frac{\alpha}{2}$, we get

$$A = a\sin\frac{N\alpha}{2}/\sin\frac{\alpha}{2}$$

$$\varphi = \angle MOX = \angle COX - \angle COM = \frac{1}{2}(\pi - \alpha) - \frac{1}{2}(\pi - N\alpha) = \frac{N-1}{2}\alpha$$

When $\alpha = 0$(same phase),we have

$$A = Na, \quad \varphi = 0$$

2. **Different Frequencies**

When two harmonic vibrations have the same direction but different frequencies, the rotating angular velocities of the two vectors differ in the rotation vector graphic method. Their phase difference is related to time, and the length and angular velocity of the compound vector change with time.

The difference between ω_1 and ω_2 is very small, furthermore $\omega_2 > \omega_1$,

$$x_1 = A_1\cos(\omega_1 t + \varphi_0), \qquad x_2 = A_2\cos(\omega_2 t + \varphi_0)$$

The synthesis of the two harmonic vibrations is

$$x = x_1 + x_2$$

$$x = 2A\cos\left(\frac{\omega_2 - \omega_1}{2}t\right)\cos\left(\frac{\omega_2 + \omega_1}{2}t + \varphi_0\right) \tag{2.36}$$

Since $\omega_1 \approx \omega_2, \ \ \omega_2 - \omega_1 \ll \omega_1$ or ω_2, we get $\frac{\omega_2+\omega_1}{2} \approx \omega_1 \approx \omega_2$.

In the combined vibration displacement expression, the first term changes slowly with time, and the second term is a harmonic function with an angular frequency close to ω_1 or ω_2. The combined vibration can be considered as a harmonic vibration with angular frequency ω and amplitude A, refer to Fig. 2.19.

The amplitude of the combined vibration changes slowly with time, and the vibration becomes strong or weak periodically, which is known as the Beat phenomenon.

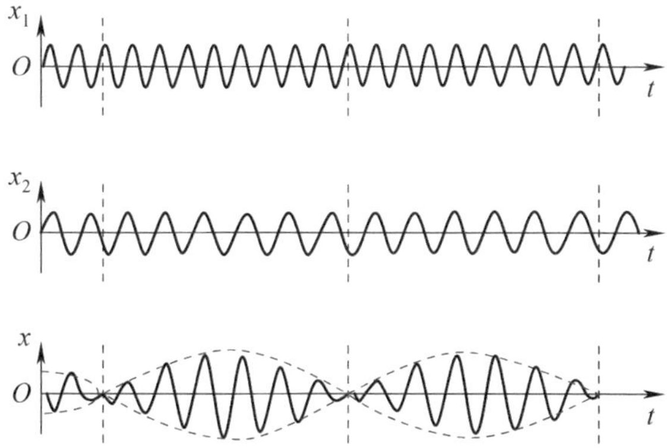

Fig. 2.19 The displacement synthesis of the two harmonic vibrations

Beat frequency: the number of times vibration changes from strong to weak in a unit of time.

$$\gamma = \left|\frac{\omega_2 - \omega_1}{2\pi}\right| = |\gamma_2 - \gamma_1| \tag{2.37}$$

2.8 Synthesis of Perpendicular Harmonic Vibrations

Expressions for the displacement of two perpendicular motions of the same frequency:

$$x = A_1 \cos(\omega t + \varphi_{10}), \quad y = A_2 \cos(\omega t + \varphi_{20})$$

take away t, we get

$$\frac{x^2}{A_1^2} + \frac{y^2}{A_2^2} - 2\frac{x}{A_1}\frac{y}{A_2}\cos(\varphi_{20} - \varphi_{10}) = \sin^2(\varphi_{20} - \varphi_{10}) \tag{2.38}$$

The combined motion is generally an ellipse in xy-plane.

After A_1 and A_2 have been determined, the ellipse's features (orientation, long or short axis, left or right screw) are determined by:

$$\Delta\varphi = \varphi_{20} - \varphi_{10} \tag{2.39}$$

Several special cases:

(1) $\varphi_{20} - \varphi_{10} = 0$

Figure 2.20 shows the two sub-vibrations in same phase, resulting in:

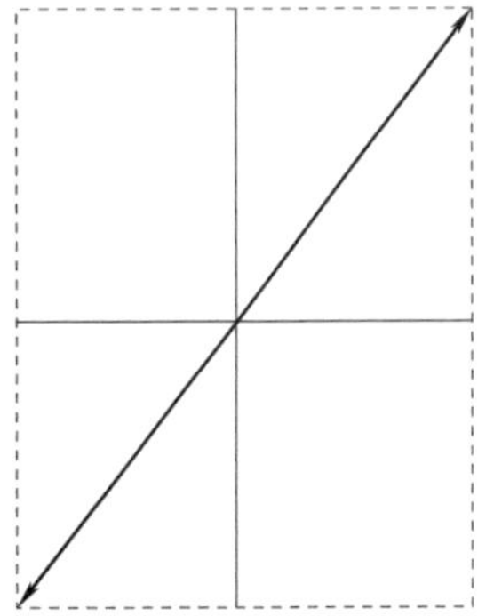

Fig. 2.20 The same phase

$$y = \frac{A_2}{A_1}x$$

The displacement from the coordination origin at any time is

$$y = \sqrt{A_1^2 + A_2^2}\cos(\omega t + \varphi) \tag{2.40}$$

(2) $\varphi_{20} - \varphi_{10} = \pi$

Figure 2.21 shows the two sub-vibrations in reversed phase, resulting in:

$$y = -\frac{A_2}{A_1}x$$

(3) $\varphi_{20} - \varphi_{10} = \pi/2$, get $\frac{x^2}{A_1^2} + \frac{y^2}{A_2^2} = 1$

Figure 2.22 illustrates an ellipse with its major axis aligned with the coordinate axis, and the particle's trajectory is clockwise.

(4) $\varphi_{20} - \varphi_{10} = 3\pi/2$, get the same expression $\frac{x^2}{A_1^2} + \frac{y^2}{A_2^2} = 1$

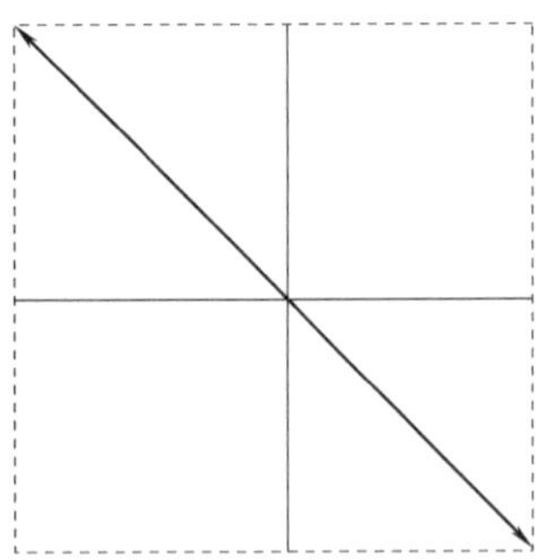

Fig. 2.21 The reversed phase

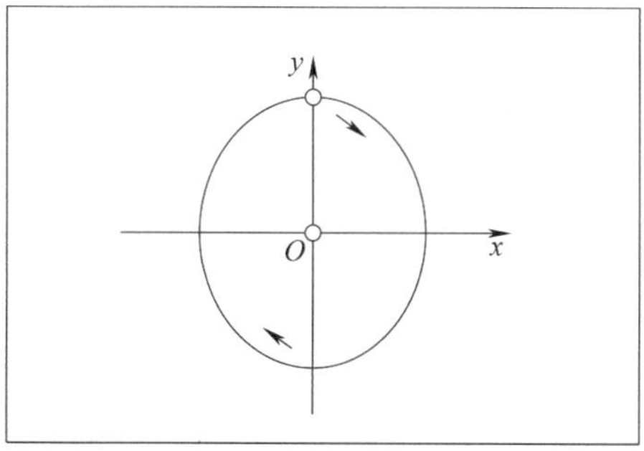

Fig. 2.22 The phase difference π/2

Figure 2.23 shows a counter clockwise particle trajectory.

Other special cases are provided in Fig. 2.24.

Synthesis of two harmonic vibrations with different frequencies and perpendicular directions results in a complex wave pattern.

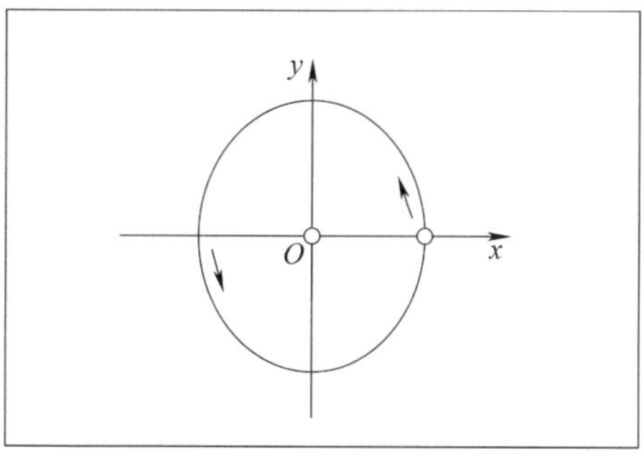

Fig. 2.23 The phase difference 3π/2

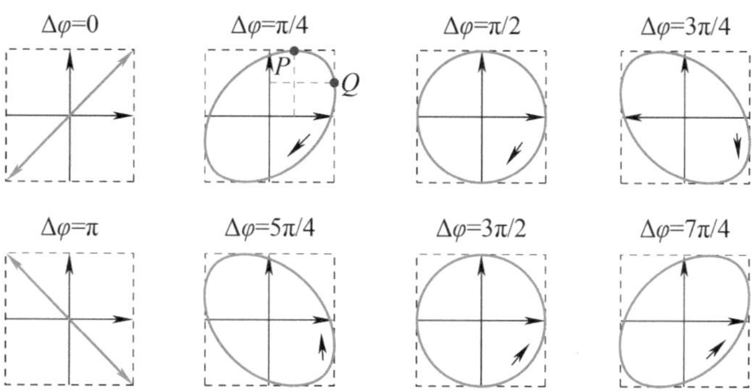

Fig. 2.24 Other cases ($\Delta\varphi = \varphi_{20} - \varphi_{10}$)

2.9 Vibration Analysis Procedure

A physical system is dynamic, with variables such as excitations (inputs) and responses (outputs) that are time-dependent. The system's response is influenced by initial conditions and excitations. Most physical systems are complex, making it impossible to incorporate all the details into a mathematical analysis. Only the most significant features are considered to predict the system's performance under given input conditions. Often, the overall performance of the system can be determined by examining even a simple model of a complex system. Vibration system analysis involves modeling, deducing, solution, and analysis of results. Analyzing a vibration system includes the following steps.

1. **Modeling**

The purpose of modeling is to represent all critical features of the system and derive mathematical (or analytical) control equations for system performance. The model should include sufficient details to describe the system using mathematical equations without excessive complexity. Depending on the system's component performance, the model can be linear or nonlinear. Linear models are quick to solve and easy to use; however, nonlinear models can reveal certain system characteristics unattainable with linear models. Therefore, a substantial amount of engineering experimentation is necessary to establish a suitable model of the vibration system.

The model is refined step by step to obtain more accurate results. Initially, a crude or simple model is used to quickly grasp the overall features of the system. Subsequently, the model is refined by incorporating more components and/or details for a clearer observation of the system's characteristics. To illustrate the refinement procedure in modeling, consider the first (end) vehicle of a train. It consists of a body, two bogies, with the bogie comprising the frame, wheelsets, and primary suspension. There are secondary suspensions between the body and the bogie's frame. The physical model of a vehicle can be simplified to 7 rigid bodies, with 23 degrees of freedom to be considered in the analysis, as shown in Fig. 2.25. If the vehicle body's mass is significantly greater than the mass of the two frames and the frame's inertia force can be ignored, the bound vibration of the vehicle body can be reduced to a single-degree-of-freedom (SDOF) vertical free vibration model, as shown in Fig. 2.26. This model can study the low-frequency bound vibration of the vehicle body. The use of simplified models facilitates beginners' understanding of railway vehicle vibration phenomena, gradually leading to an understanding of complex models. The simplified SDOF vehicle vibration model can describe the main characteristics of vehicle vertical vibration, and its vertical free vibration equation at the static equilibrium position is:

$$\boldsymbol{M}\ddot{\boldsymbol{Z}} + \boldsymbol{C}\dot{\boldsymbol{Z}} + \boldsymbol{K}\boldsymbol{Z} = \boldsymbol{0} \tag{2.41}$$

where, $\boldsymbol{C}$ is the vertical damper coefficient matrix, $\boldsymbol{K}$ is the vertical stiffness matrix, and $\boldsymbol{M}$ is the mass matrix of the vehicle.

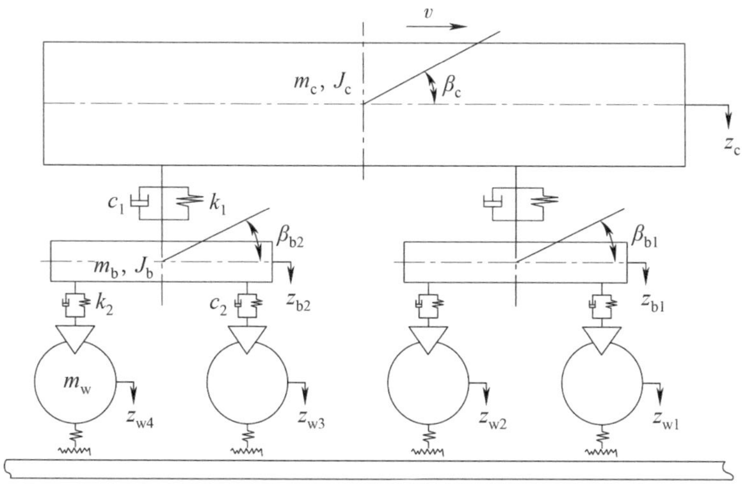

Fig. 2.25 The modeling diagram of a railway vehicle

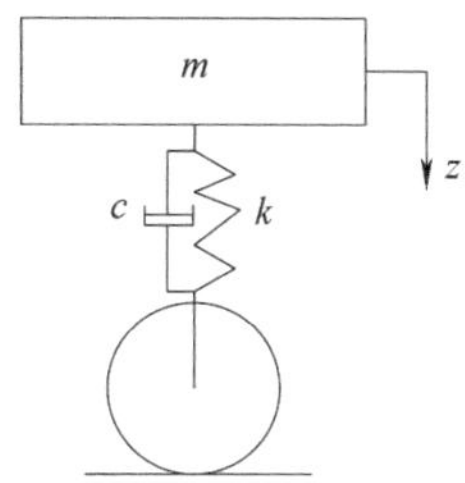

Fig. 2.26 SDOF model of a simplified vehicle

2. **Deducing the Equations**

Once the model is available, use the principles of dynamics to deduce the system's equations of motion. The equations of motion can be conveniently deduce by drawing freebody diagram of all involved masses. The equations of motion are usually in the form of a set of ordinary differential equations (ODE) for a discrete system and partial differential equations for a continuous system. The equations may be linear or nonlinear, depending on the system's components'characteristics. Several approaches are commonly used to deduce the equations, including Newton's second law of motion, d′ Alembert principle, and the principle of conservation of energy.

3. **Solution of the Equations**

The equations of motion must be solved to find the system's response. The following techniques can be used to find the solution: standard methods of solving differential equations, Laplace transform methods, matrix methods, and numerical methods. If the system is nonlinear, it can seldom be solved in closed form. Furthermore, solving partial differential equations is far more complicated than that of ODE. Numerical methods can be employed to solve the equations.

4. **Analysis of Results**

The solution of the equations provides the displacements, velocities, and accelerations of the system's masses. These results must be interpreted with a clear understanding of the analysis's purpose and the potential design implications of the results.

Exercise 2

E2.1: Consider the forging hammer shown in Fig. 2.27. It consists of a frame, a falling weight known as the tup, an anvil, and a foundation block. Model the forging hammer.

E2.2: A cantilever beam carries a mass M at the free end as shown in Fig. 2.28. A mass m falls from a height h onto the mass M and adheres to it without rebounding. Determine the resulting transverse vibration of the beam.

E2.3: A simply supported beam of square cross-section and length of 1 m, carrying a mass of 2.3 kg at the middle, is found to have a natural frequency of transverse vibration of 30 rad/s. Determine the Young's modulus of elasticity of the beam.

E2.4: Any rigid body pivoted at a point other than its center of mass will oscillate about the pivot point under its own gravitational force. Such a system is known as a compound pendulum (Fig. 2.29). Find the natural frequency of such a system.

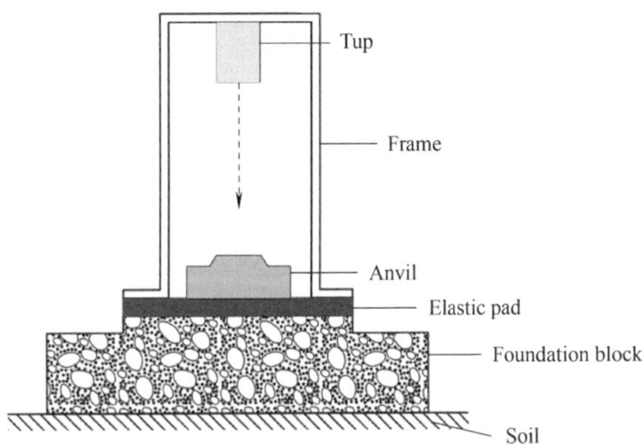

Fig. 2.27 Forging hammer model

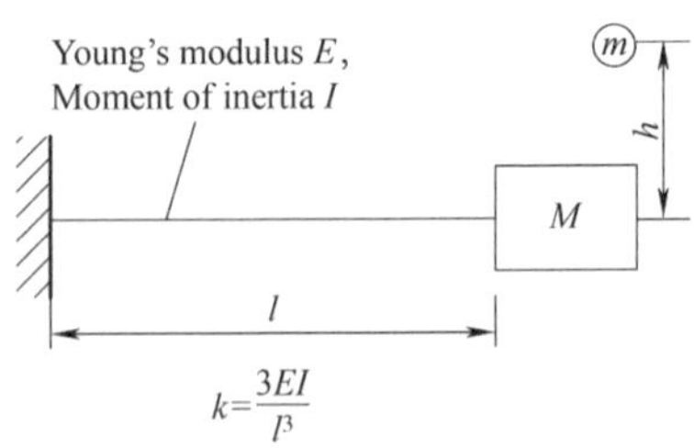

Fig. 2.28 A cantilever beam system

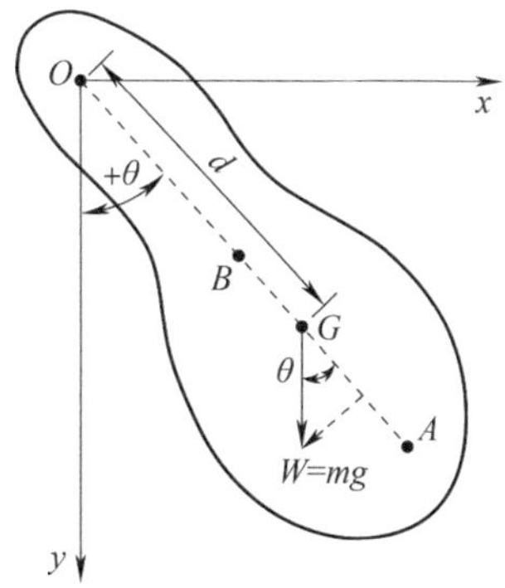

Fig. 2.29 The compound pendulum

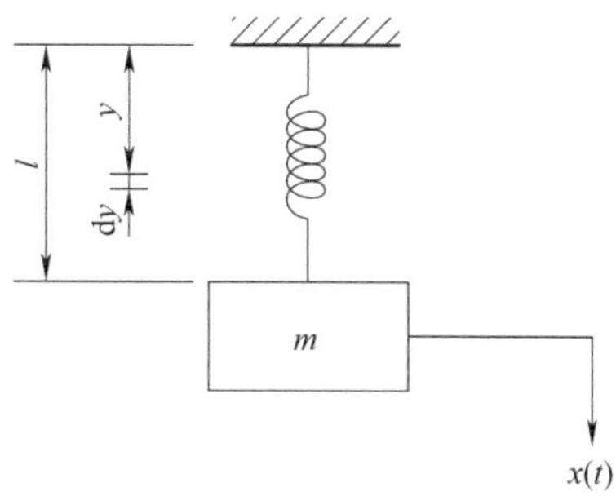

Fig. 2.30 The equivalent mass of a spring

E2.5: Determine the effect of the mass of the spring on the natural frequency of the spring- mass system shown in Fig. 2.30.

Chapter 3
Random Variable

3.1 Introduction to Stochastic Analysis

3.1.1 Random Vibration in the Real-World

The real-world vibration environment is inherently random. These vibrations are commonly encountered in various contexts:

(1) On railway vehicles, such as track irregularities and aerodynamic turbulent flow around the vehicles'surface, necessitating the analysis of railway vehicles' response to random track excitations in both time and frequency domains.
(2) On road vehicles, the irregularities of the roads.
(3) On aircraft, engine noise and aerodynamic turbulent flow around the wings and fuselage, creating non-stationary pressures, etc.
(4) On ships, engine noise and swell.
(5) On missiles, the majority of vibrations encountered by military equipment, particularly the internal components of guided missiles, are random with respect to time and exhibit a continuous spectrum: the gas jet emitted at high velocity creates significant turbulences, resulting in acoustic noise that impacts the missile's skin until exceeding certain speed (or until it leaves the Earth's atmosphere).
(6) In many other real-life scenarios, etc.

Random vibration tests first emerged around 1955 due to the limitations of sine tests in exciting multiple resonances. The trend is to replace swept sine tests, which excites resonances, with random vibration tests whose effects more closely mimic the real environment. Random vibration testing is a more pragmatic approach:

① To identify the structure (studying resonance frequency and measuring the *Q*-factor),the advantage is that testing time is reduced.

J. Zhang and H. Zhou, *Random Vibration and Spectrum Analysis*,
https://doi.org/10.1007/978-981-95-0191-5_3

② To simulate the effects of high-frequency shocks that are difficult to achieve with simple forms.

3.1.2 Uncertainties in Structural Engineering Problems

In the field of engineering, the design, processing, and application performance of any system will exhibit variability, and complete control in practical research is unattainable. Any system and process will inevitably contain uncertainties, leading to random characteristics in structural properties, geometric features, and system traits, thereby influencing system performance. Consequently, in engineering systems, accounting for the random characteristics of parameters and employing stochastic structural models to analyze structural design performance is more effective in capturing essential issues than traditional models. The following factors are primarily considered in the engineering domain:

(1) Loads (earthquakes, wind, waves, guideway unevenness, etc.).
(2) Structural properties (elastic constants, mass, damping, strength, boundary conditions, joints, etc.).
(3) Modeling (analytical, computational, and experimental).
(4) Condition assessment of existing structures.
(5) Human errors.

3.1.3 Stochastic Dynamics

Stochastic dynamics are mathematical methodologies for examining the evolution of various systems over time, which research scope encompass everything from microscopic particles like molecules, atoms, and particles to the motion of celestial bodies in the universe. The approach of stochastic dynamics involves establishing mathematical models to describe the motion laws and evolution processes of physical systems.

The methodology of stochastic dynamics involves characterizing the motion laws of physical systems through stochastic differential equations. A stochastic differential equation is a type of differential equation that includes random terms and can describe the random motion of a physical system. Solving stochastic differential equations requires the application of stochastic process theory and numerical computation techniques.

In the engineering field, stochastic dynamics primarily focuses on the following aspects:

(1) A branch of dynamics where uncertainties in loads are quantified using probability theory, random processes, and statistics.

(2) Random vibration analysis and probabilistic structural dynamics.
(3) Structural failure under uncertain dynamic loads.
(4) Design of structures under uncertain dynamic loads.

3.1.4 Mathematical Models for Uncertainty

Uncertainty modeling refers to the process of translating uncertain factors into mathematical models for analysis and computation. During this process, it is necessary to convert uncertainty factors into probability distributions and estimate the parameters of these distributions. Commonly utilized probability distributions include the normal distribution, Poisson distribution, exponential distribution, etc. It is also essential to model the relationships between uncertain factors to analyze complex uncertain systems. Common methodologies include Bayesian networks, Markov chains, etc.

Uncertainty modeling and computation have extensive applications across various fields. In medicine, uncertainty modeling and computation aids in decision-making regarding patient diagnosis and treatment. In natural disaster and environmental management, it predicts and responds to natural disaster risks. In machine learning and artificial intelligence, uncertainty modeling and computation hold significant value.

The common uncertainty modeling methods in engineering primarily adopt the following principles:

(1) Probability, random variables, random processes, and statistics.
(2) Fuzzy logic.
(3) Interval algebra.
(4) Convex models.

3.1.5 Definition of Probability

Probability, reflects the likelihood of a random event occurring. Random events are those that may or may not occur under identical conditions. There are four kinds of definitions: classical definition, relative frequency definition, statistical definition, and axiomatic definition.

1. **Classical Definition**

If an experiment meets two conditions:

(1) There are a finite number of basic outcomes in the experiment.
(2) The likelihood of each basic outcome occurring is the same.

Such an experiment is considered a classical experiment.

For event A in a classical experiment, its probability is defined as $P(A) = m/n$, where n represents the total number of possible basic outcomes in the experiment,

n represents the number of basic experimental outcomes included in event A. This method of defining probability is known as the classical definition of probability.

For example: P (getting an even number on a die toss) $= 3/6 = 1/2$.

Objections:

① What is "equal likelihood"?
② What if not "equal likelihood"? (What is the probability that the sun will rise tomorrow?)
③ No room for experimentation.
④ Probability must be a rational number.

2. **Relative Frequency**

As problem complexity increases, the vulnerability of equal likelihood becomes evident, especially for the same event, different probabilities can be calculated from different perspectives of equal likelihood, leading to various paradoxes. On the other hand, with accumulating experience, people realize that when conducting a large number of repeated experiments, the frequency of an event consistently oscillates around a fixed number, indicating a degree of stability. R. Von Mises defines this fixed number as the probability of the event, which is the frequency definition of probability. In theory, the frequency definition of probability lacks rigor.

3. **Statistical Definition**

Under certain conditions, repeat n experiments, where n_A is the number of times event A occurs in n experiments. If the frequency n_A/n gradually stabilizes near a certain value p as n gradually increases, then the value p is called the probability of event A occurring under that condition, denoted as $P(A) = p$. This definition is known as the statistical definition of probability.

Since the frequency always lies between 0 and 1, according to the statistical definition of probability, for any event A, there are $0 \leq P(A) \leq 1$, $P(\Omega) = 1$, $P(\Phi) = 0$, where Ω, Φ represent inevitable events (events that certain occur under certain conditions) and impossible events(events that cannot occur under certain conditions), respectively.

4. **Axiomatic Defination**

Let E be a random experiment and S be its sample space. Assign a real number to each event A of E, denoted as $P(A)$, and call it the probability of event A. Here $P(A)$ is a set function that satisfies the following conditions:

(1) Non-negativity: for each event A, $P(A) \geq 0$.
(2) Normality: for inevitable events, $P(\Omega) = 1$.
(3) Countable additivity: Let $A_1, A_2, \ldots,$ be mutually incompatible events, that is, for $i \neq j, A_i \cap A_j = \Phi$, $(i, j = 1, 2, \ldots)$, then $P(A_1 \cup A_2 \cup \cdots) = P(A_1) + P(A_2) + \cdots$.

Sample space Ω : set of all possible outcomes of a random experiment.

3.1.6 Properties of Probability

Probability has eight distinct properties:

(1) For any event A,

$$P(A) \leq 1 \tag{3.1}$$

(2) When A_C is the complement of A,

$$P(A_C) = 1 - P(A) \tag{3.2}$$

(3) When events A and B satisfy that A is included in $B(A \subset B)$,

$$P(B - A) = P(B) - P(A), \quad P(A) \leq P(B) \tag{3.3}$$

(4) For any two events A and B,

$$P(A \cup B) = P(A) + P(B) - P(A \cap B) \tag{3.4}$$

(5) When n events $A_1, A_2, \ldots, A_n$ are mutually incompatible(finite additivity),

$$P(A_1 \cup A_2 \cdots \cup A_n) = P(A_1) + P(A_2) \cdots + P(A_n) \tag{3.5}$$

(6) For any two events A and B,

$$P(B - A) = P(B) - P(A \cap B) \tag{3.6}$$

(7) Inevitable events

$$P(\Phi) = 0 \tag{3.7}$$

(8) Impossible events

$$P(\Omega) = 1 \tag{3.8}$$

3.1.7 Conditional Probability and Stochastic Independence

1. **Conditional Probability**

When addressing many probability problems, it is often necessary to calculate the probability of events with certain additional information (conditions). If we calculate

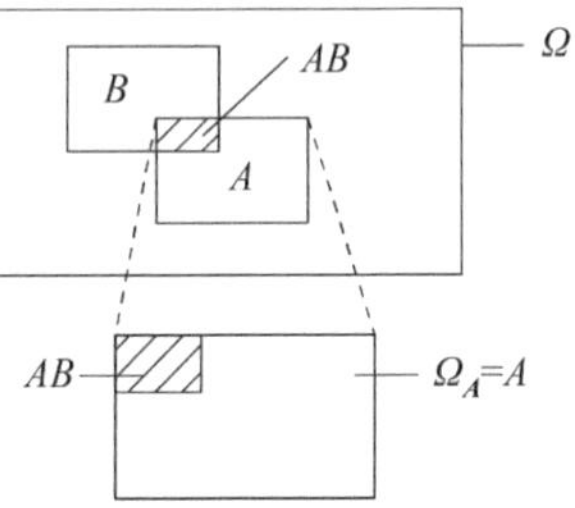

Fig. 3.1 Probability of event A given that B

the probability of event B occurring given that event A has occurred, we record this probability as $P(A|B)$.

Generally, $P(A|B) \neq P(A).P(A|B) =$ Probability of event A given that B has occurred, as shown in Fig. 3.1.

Example 3.1 Fair die tossing $\Omega = \{1, 2, 3, 4, 5, 6\}$, the die has been tossed, and an even number has been observed, P(2|Even) = ?

Solution:

Approach 1: $Even = (2 \quad 4 \quad 6)$; $P(2|Even) = 1/3$ (classical definition)

Approach 2:

$$P(2|Even) = \frac{P\{2 \cap Even\}}{P(Even)}$$

$$\{2 \cap Even\} = \{2\}$$

$$P(2|Even) = \frac{1/6}{1/2} = 1/3$$

2. **Stochastic Independence**

Events A and B are said to be stochastically independent if any one of the following four statements is true.

(1) The probability of occurrence of event A is unaffected by the occurrence of event B.
(2) $P(A \cap B) = P(A)P(B)$
(3) $P(A|B) = P(A)$
(4) $\frac{P(A \cap B)}{P(B)} = P(A), \quad P(B) \neq 0$

Notation: events A and B are independent.

$$P(A \perp B) \Rightarrow P(A \cap B) = P(A)P(B) \tag{3.9}$$

Remarks Definition (1) is not useful for verifying if events A and B are independent.

If we need to verify if events A and B are independent, we need to find $P(A)$, $P(B)$, $P(A \cap B)$, and use the definitions.

Independence of more than two events can also be defined. Thus

$$P(A_i \cap A_j) = P(A_i)P(A_j), \quad i, j = 1, 2, 3, \ldots (i \neq j).$$
$$P(A_1 \cap A_2 \cap A_3 \cdots) = P(A_1)P(A_2)P(A_3) \cdots.$$

3.2 Fundamental Operations of Events

There are two events A and B, with the sample space being Ω, as shown in Fig. 3.2. The fundamental operations are as follows.

(1) $P(\overline{A}) = 1 - P(A)$

(2) $P(A \cup B) = P(A) + P(B) - P(AB)$,

if event A, B mutually exclusive, $P(A \cup B) = P(A) + P(B)$.

(3) $P(A - B) = P(A \cap \overline{B}) = P(A) - P(AB)$, if $B \subseteq A$, $P(A - B) = P(A) - P(B)$

(4) $$\begin{aligned} P(AB) &= P(A)P(B|A), \quad P(A) > 0 \\ &= P(B)P(A|B), \quad P(B) > 0 \end{aligned}$$

if event A, B mutually independent, $P(AB) = P(A)P(B)$.

(5) $P(B|A) = P(AB) \div P(A), \quad P(A) > 0$.

Example 3.2 Toss two coins, resulting in two events: Head and Tail. And the sample space is Ω =($hh\ ht\ th\ tt$). Let $a, b \geq 0$, such $that(a + b) = 1$. Let $P(hh) = a^2, P(tt) = b^2, P(th) = P(ht) = ab$.

Clearly, $P(\Omega) = P(hh) + P(tt) + 2P(th) = (a + b)^2 = 1$.

Define two events: E_1 = head on the first toss = $\{hh \quad ht\}$, $\quad E_2$ = head on the second $toss$ =$\{hh\ ht\}$. Question: verify if E_1 and E_2 are independent.

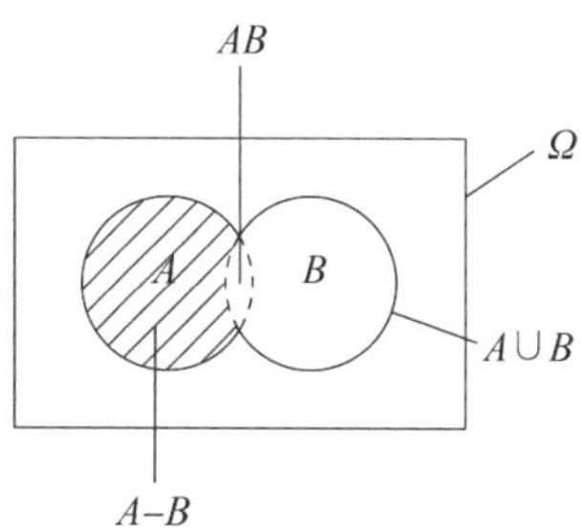

Fig. 3.2 Event A and B in the sample space

Proof:

$$P(E_1) = P(hh, ht) = a^2 + ab = a(a + b) = a$$

$$P(E_2) = P(hh, th) = a^2 + ba = a(a + b) = a$$

$$P(E_2 \cap E_2) = P(hh) = a^2$$

If E_1 and E_2 are independent,

$$P(E_1 \cap E_2) = P(E_1)P(E_2)$$

Example 3.3 In which three events are pairwise (each other) independent but are not independent as a group. Consider a fair tetrahedron (this has four faces). Let the four faces be painted as: Green (G), Yellow (Y), Black (B), and $G + Y + B$. Determine if G, Y, B, and $G + Y + B$ are independent.

Illustration:

$$P(G) = P(B) = P(Y) = \frac{1}{4} + \frac{1}{4} = \frac{1}{2}$$

$$P(GY) = \frac{1}{4} = P(G)P(Y)$$

$$P(YB) = \frac{1}{4} = P(Y)P(B)$$

$$P(GYB) = \frac{1}{4} \neq P(G)P(Y)P(B) = \frac{1}{8}$$

3.2.1 Chain Rule

$$P(A_1A_2 \cdots A_n) = P(A_1)P(A_2|A_1)P(A_3|A_1A_2) \cdots = P(A_n|A_1A_2 \cdots A_{n-1}) \quad (3.10)$$

Figure 3.3 shows a complete event group and segmentation, and

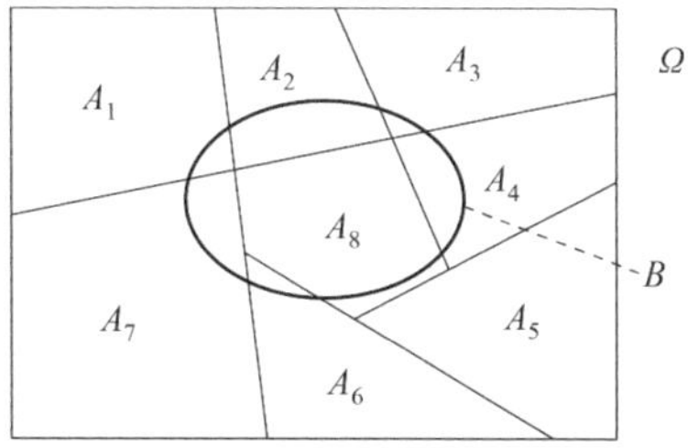

Fig. 3.3 The complete event group and segmentation

$$A_i(i = 1, 2, \ldots, n), meet : \forall_i \neq j, \quad A_iA_j = \phi, \quad \bigcup_{i=1}^{n} A_i = \Omega. \tag{3.11}$$

3.2.2 Total Probability Theorem

Let $\{A_i\}_{i=1}^{N}$ constitute a particular subset of Ω. That is

$$\bigcup_{i=1}^{N} A_i = \Omega, \quad A_i \cap A_j = \Phi, \quad \forall i \neq j \tag{3.12}$$

Let B be a set

$$B = \bigcup_{i=1}^{N} (A_i \cap B) \tag{3.13}$$

$$P(B) = P\left(\bigcup_{i=1}^{N} (A_i \cap B)\right) = \sum_{i=1}^{N} P(A_i \cap B) \tag{3.14}$$

$$P(B) = \sum_{i=1}^{N} P(B|A_i)P(A_i) \tag{3.15}$$

Bayes' theorem

$$P(A_i|B) = \frac{P(A_i \cap B)}{P(B)} = \frac{P(B|A_i)P(A_i)}{P(B)} \tag{3.16}$$

$$P(A_i|B) = \frac{P(B|A_i)P(A_i)}{\sum_{i=1}^{N} P(B|A_i)P(A_i)} \tag{3.17}$$

where, $P(A_i)$ is a priori probability, $P(A_i|B)$ is a posteriori probability.

3.3 Probability Distribution Function

In practical problems, it is often necessary to study the probability of a random variable ξ being less than a given value x. This is a function of x. The function is known as the distribution function of the random variable ξ, denoted as $F(x)$, where $F(x) = P(\xi < x)$ $(-\infty < x < +\infty)$. It can determine the probability of a random variable falling within any range. For instance, in bridge and dam design, the probability of the river's highest water level being less than x meters is a function of every year. This function is the distribution function of the highest water level ξ.

3.3.1 *Probability Distribution Function of $P_X(x)$*

The probability distribution function is a mathematical representation that describes the distribution of random variable values. For any real number x, the probability of event $(X \leq x)$ is naturally a function of x, given

$$F(x) = P(X \leq x) \tag{3.18}$$

Clearly,

$$F(-\infty) = 0, \quad F(\infty) = 1 \tag{3.19}$$

$F(x)$ is called the distribution function of the random variable X. Thus, the distribution function $F(x)$ completely determines the probability of an event $(a \leq X \leq b)$, in other words, the distribution function $F(x)$ fully describes the statistical characteristics of the random variable X.

3.3.2 *Basic Properties of the Probability Distribution Function*

The distribution function $F(x)$ is an ordinary function. It is through it that random variables can be studied using mathematical analysis methods.If X is considered as the coordinate of a random point on the number axis, then the value of the distribution function $F(x)$ at x represents the probability of X falling into the interval.

The distribution function $F(x)$ has the following basic properties:

(1) $F(-\infty) = 0, \quad F(+\infty) = 1.$
(2) $F(x)$ is a monotone non-decreasing function and is right continuous, namely $F(x_1) \leq F(x_2)$ when $x_1 < x_2, \quad F(x^+) = F(x).$
(3) Probability formulas of section events

$$P\{x_1 < x < x_2\} = F(x_2) - F(x_1) \tag{3.20}$$

$$P\{x_1 \leq x \leq x_2\} = P\{x_1 < x \leq x_2\} + P(x = x_1) \tag{3.21}$$

$$P\{x_1 < x < x_2\} = P\{x_1 < x \leq x_2\} - P(x = x_2) \tag{3.22}$$

$$P(X = x) = F(x) - F(x^-). \tag{3.23}$$

3.3.3 Random Variables

1. **Probability Distribution of Discrete Random Variables**

For discrete random variables, if they are the values of the variable X and the probability is corresponding to the above value, then the probability distribution of the discrete random variable X is

$$P(X = x_i) = p_i, \quad i = 1, 2, \ldots, n \tag{3.24}$$

And the probability p_i should meet the criteria $\sum_{i=1}^{n} p_i = 1$. Therefore, the probability distribution function of the discrete random variable X is

$$F(x) = \sum_{x_i < n} p_i. \tag{3.25}$$

2. **Probability Distribution of Continuous Random Variables**

For continuous random variables, let the variable X take value in the interval (a, b), and its distribution function $F(x)$ is a monotonic increasing function, and it is differentiable and its derivative $F'(x)$ is continuous in $-\infty < x < \infty$, then the probability of the variable X falling between x and $x + \Delta x$ is

$$P(x \leq X \leq x + \Delta x) = F(x + \Delta x) - F(x) \tag{3.26}$$

To describe its probability distribution law, it is not possible to use distribution columns, but to introduce a new concept of "probability distribution density function". Define the derivative $F'(x)$ of the probability distribution function $F(x)$ as the probability distribution density function $f(x)$, i. e.,

$$f(x) = F'(x) = \lim_{\Delta x \to 0} \frac{F(x + \Delta x) - F(x)}{\Delta x} \tag{3.27}$$

The probability distribution function of the continuous random variable X can be written as:

$$F(x) = \int_{-\infty}^{x} f(x)\mathrm{d}x \tag{3.28}$$

As long as the probability distribution density function $f(x)$ of a continuous random variable X is known, the probability of X falling within a certain interval (x_1, x_2) can be obtained:

$$P(x_1 \leq X \leq x_2) = F(x_2) - F(x_1) = \int_{x_1}^{x_2} f(x)\mathrm{d}x \tag{3.29}$$

Similar to the probability function of discrete random variables, for the distribution density function

$$\int_{x_1}^{x_2} f(x)\mathrm{d}x = 1, \quad f(x) \geq 0 \tag{3.30}$$

3. **Mixed Probability Distribution**

In practical applications, it is often encountered that multiple probability distributions are superimposed or mixed with each other, which is called a mixed probability distribution. The mixed probability distribution can more accurately describe complex random phenomena in the real world and has a wide range of applications. A mixed probability distribution is a composite probability distribution formed by a linear combination of multiple basic probability distributions. It contains multiple sub-distributions, each with its own weights and parameters. Essentially, a hybrid model can be seen as a "model within a model".

For continuous random variables, the mixed probability density function $f(x)$ can be expressed as a linear combination of each sub-distribution density function $g\_i(x)$ and its corresponding weight w_i.

$$f(x) = w\_1 \times g\_1(x) + w\_2 \times g\_2(x) + \cdots + w\_n \times g\_n(x) \tag{3.31}$$

where, w_i represents the weight of the i-th sub-distribution, satisfying $0 \leq w\_i \leq 1$ and $\sum w\_i = 1$; $g\_i(x)$ is the probability density function of the i-th sub-distribution.

The mixed probability distribution has the following important properties:

(1) Variable Morphology

The mixed probability distribution can adapt to different forms of data distribution based on different combinations of weights and parameters. This enables the hybrid model to better fit complex data distributions.

(2) Nonlinear Relationships

Since the mixed probability distribution is a linear combination of multiple subdistributions, it has a nonlinear relationship. This enables the hybrid model to describe nonlinear related random phenomena.

(3) Diversity

A mixed probability distribution can contain multiple sub-distributions of different types, which can better describe diverse random phenomena. For example, a mixed Gaussian model can fit multiple Gaussian distributions simultaneously.

3.3.4 Probability Density Function (PDF)

The probability density function of a continuous random variable (the density function when not confusing) is a function that describes the likelihood of the output value of this random variable near a certain value point. The probability that the value of a random variable falls within a certain region is the integral of the probability density function in that region. When the probability density function exists, the cumulative distribution function is the integral of the probability density function. The probability density function is generally marked in lowercase.

The density probability function $P(x)$ has the following basic properties:

(1) Definition

$$p_X(x) = \frac{\mathrm{d}P_X(x)}{\mathrm{d}x} \Rightarrow P_X(x) = \int_{-\infty}^{x} p_X(u)\mathrm{d}u$$

(2) Properties

$$p_X(\infty) = P\{X \le \infty\} = P(\Omega) = \int_{-\infty}^{\infty} p_X(u)\mathrm{d}u$$

$$P\{a \le X \le b\} = \int_{a}^{b} p_X(u)\mathrm{d}u$$

$$p_X(x)\mathrm{d}x = P\{x \le X \le x + \mathrm{d}x\}$$

(3) $0 \le P_X(x) \le 1$.
(4) $P_X(\infty) = P_X\{X \le \infty\} = P(\Omega) = 1$.
(5) $P_X(-\infty) = P_X\{X \le -\infty\} = P(\varphi) = 0$.
(6) $x_2 > x_1 \Rightarrow P_X(x_2) \ge P_X(x_1)$, PDF is monotone nondecreasing.
(7) $\lim_{0<\varepsilon\to 0} P_X(x+\varepsilon) = P_X(x)$, PDF is the right continuous.

The discrete probability density distribution function is a function used to describe the probability distribution of discrete random variables. It can be used to calculate the probability of a random variable taking a specific value, as well as the probability of a random variable taking a value within a certain interval. Some important properties of discrete probability density distribution functions are as follows. Firstly, its value must be between 0 and 1, as the range of probability values is 0 to 1. Secondly, the sum of probabilities for all possible values must be equal to 1, as a random variable must take a value. Finally, the discrete probability density distribution function can be used to calculate the expected values and variances of random variables.

Furthermore, $P_X(X = k) = p_k(k = 0, 1, 2, \ldots)$ is also known as the probability mass function (PMF), and PDF is also known as the cumulative probability distribution function (CPDF).

3.3.5 Expression for Several Special Functions

This section will introduce the expressions of rectangular functions, Dirac's delta functions, and staircase functions based on step functions, and then introduce the probability density of Bernoulli random variables and their applications.

(1) Heaviside's Step Function

$$U(x-a) = \begin{cases} 0 & x < a \\ 1 & x > a \\ 1/2 & x = a \end{cases} \tag{3.32}$$

(2) Box Function

$$f(x) = U(x-a) - U(x-b) \tag{3.33}$$

(3) Dirac's Delta Function

$$f(x) = F_0[U(x) - U(x-a)] \tag{3.34}$$

$$\lim_{\substack{a\to 0\\ F_0 a\to 1}} f(x) = \lim_{\substack{a\to 0\\ F_0 a\to 1}} F_0[U(x) - U(x-a)]$$
$$= \lim_{\substack{a\to 0\\ F_0 a\to 1}} a\frac{[U(x) - U(x-a)]}{a} = \frac{d}{dx}U(x) = \delta(x) \tag{3.35}$$

$$\begin{cases} \delta(x-a) = 0, \quad x \neq a \\ \int_{-\infty}^{\infty} \delta(x-a)\mathrm{d}x = 1 \\ \int_{-\infty}^{\infty} f(x)\delta(x-a)\mathrm{d}x = f(a) \\ \frac{\mathrm{d}}{\mathrm{d}x}U(x-a) = \delta(x-a) \end{cases} \tag{3.36}$$

(4) Staircase Function

$$f(x) = \alpha[U(x) - U(x-a)] + \beta[U(x-a) - U(x-b)] \tag{3.37}$$

$$\frac{\mathrm{d}f}{\mathrm{d}x} = \alpha[\delta(x) - \delta(x-a)] + \beta[\delta(x-a) - \delta(x-b)]. \tag{3.38}$$

3.3.6 *Bernoulli Trials and Binomial Random Variable*

Commonly encountered random variables:

(1) Models for rare events.
(2) Models for sums.
(3) Models for products.
(4) Models for extremes: highest and lowest.
(5) Models for waiting times. The Limit theorems are introduced, and Binomial random variables are used to solve the above application problems.

The random experiment has only two outcomes: success S and failure F.

$$\Omega = \{S \quad F\} \tag{3.39}$$

Let

$$P(S) = P(X = 0) = p \tag{3.40}$$

$$P_X(x) = pU(x) + (1-p)U(x-1) \tag{3.41}$$

$$p_X(x) = p\delta(x) + (1-p)\delta(x-1) \tag{3.42}$$

Check

$$\int_{-\infty}^{\infty} p_X(x)dx = \int_{-\infty}^{\infty} \left[p\delta(x) + (1-p)\delta(x-1)\right]dx = 1 \tag{3.43}$$

From Eqs. (3.39) to (3.43), we notice that:

① p is the parameter of the Bernoulli random variable.
② Discrete random variable.
③ Finite sample space.
④ Basic building block.

If the trial satisfies the four items, we call it a Bernoulli trial.

Repeated Bernoulli trials result in a Binomial random variable. The random experiment here consists of repeated Bernoulli trials.

Assumptions

① The random experiment consists of N independent trials.
② Each trial results in only two outcomes (success/failure).
③ P(success) remains constant during all trials.

Define X is the number of successes in N trials, $X = 0, 1, 2, 3, \ldots, N$.

$$P(X = k) = C_k^N p^k (1-p)^{N-k}, \quad k = 0, 1, 2, \ldots, N \tag{3.44}$$

Consider a sequence of N trials resulting in k successes. The occurrence of successes implies the occurrence of $(N-k)$ failures. The probability of occurrence of one such sequence is $p^k(1-p)^{N-k}$. The number of such possible sequences is C_N^k. These sequences are mutually exclusive. Get

$$P(X = k) = C_N^k p^k (1-p)^{N-k}$$

$$\Rightarrow P(X \le m) = \sum_{k=0}^{m} C_N^k p^k (1-p)^{N-k} \tag{3.45}$$

$$\Rightarrow P(X \le N) = \sum_{k=0}^{N} C_N^k p^k (1-p)^{N-k} = 1 \tag{3.46}$$

3.3.7 Geometric Random Variable

In the Bernoulli experiment, the probability of an event A occurring in each experiment is recorded as p, and the experiment stops when the event A occurs. At this time, the number of experiments conducted is N. Random experiments as in binomial random variables. N is number of trials for the first success, $N = 1, 2, 3, \ldots$.

P (first success in N-th trial) $= P$ [success on the N-th trial, failures on the first $(N-1)$ trial]

$$\Rightarrow P(N = n) = (1-p)^{n-1}p, \quad N = 1, 2, 3, \ldots \tag{3.47}$$

$$\sum_{n=1}^{\infty} P(N = n) = \sum_{n=1}^{\infty} (1-p)^{n-1}p = 1 \tag{3.48}$$

For example: let $p = 0.6$

$$\Rightarrow \sum_{n=1}^{\infty} (1-p)^{n-1}p = \sum_{n=1}^{\infty} 0.4^{n-1} \times 0.6 = 0.6\left(1 + 0.4 + 0.4^2 + 0.4^3 + \cdots\right)$$

The expression inside the bracket is a geometric progression. Hence the name geometric random variable.

For a geometric random variable with $p = 0.4$, it is a pascal or negative binomial distribution.

Define W_k is the number of trials to the first success, it can be shown that

$$P(W_k = w) = C_{w-1}^{k-1}(1-p)^{w-k}p^k, \quad w = k, k+1, k+2, \ldots, \infty. \tag{3.49}$$

3.4 Poisson PDF as a Limit of Binomial PDF

The Poisson distribution is a discrete probability distribution commonly encountered in statistics and probability, published by French mathematician Siméon Denis Poisson in 1838.

The probability function of the Poisson distribution is

$$P(X = k) = \mathrm{e}^{-\lambda}\frac{\lambda^k}{k!}, \quad k = 0, 1, 2, \ldots \tag{3.50}$$

The parameter λ represents the average number of occurrences of random events per unit time (or area). The Poisson distribution is suitable for describing the number of times a random event occurs per unit time. The expectation and variance of the Poisson distribution are λ.

The characteristic function of the Poisson distribution is

$$\phi(t) = \exp\left\{\lambda(\mathrm{e}^{it} - 1)\right\} \tag{3.51}$$

The Poisson distribution is suitable for describing the number of times a random event occurs per unit time (or space). For example, the number of people arriving at a certain service facility within a certain period of time, the number of calls received by the telephone exchange, the number of waiting passengers at the bus station, the number of machine failures, the number of natural disasters, the number of defects on a product, the distribution of bacteria in a unit partition under a microscope, and so on.

A PMF of $B(n, p)$ is given by

$$P(X = k) = C_n^k p^k (1-p)^{n-k}, \quad k = 0, 1, 2, \ldots, n \tag{3.52}$$

Let $n \to \infty, \quad p \to 0, \quad k \ll n$, such that $np \to a$ (rare events)

$$C_n^k p^k (1-p)^{n-k} \to \mathrm{e}^{-a} \frac{a^k}{k!}. \tag{3.53}$$

3.5 Gaussian (Normal) Random Variable Model for Sums

The Gaussian distribution, also known as the normal distribution, is a commonly used continuous probability distribution. The PDF of a Gaussian distribution is a two-dimensional function, where one dimension is the value of a random variable, and the other dimension is the probability density of the random variable.

The probability density function of the Gaussian distribution is:

$$p_X(x) = \frac{1}{\sqrt{2\pi}\sigma} \exp\left[-\frac{(x-m)^2}{2\sigma^2}\right], \quad -\infty < x < \infty \tag{3.54}$$

where m and σ are the parameters of a Gaussian variable, $-\infty < m < \infty, \quad \sigma \geq 0$. Gaussian random variable is denoted by $N(m, \sigma)$.

$$N(m, \sigma) \Rightarrow p_X(x) = \frac{1}{\sqrt{2\pi}} \exp\left[-\frac{1}{2}\left(\frac{x-m}{\sigma}\right)^2\right], \quad -\infty < x < \infty \tag{3.55}$$

$$\int_{-\infty}^{\infty} p_X(x)\mathrm{d}x = 1$$

$$\int_{-\infty}^{\infty} p_X(x)\mathrm{d}x = \frac{1}{\sqrt{2\pi}\sigma} \int_{-\infty}^{\infty} \exp\left[-\frac{1}{2}\left(\frac{x-m}{\sigma}\right)^2\right]\mathrm{d}x$$

Substitute, $u = \left(\frac{x-m}{\sigma}\right) \Rightarrow \sigma\mathrm{d}u = \mathrm{d}x \Rightarrow \int_{-\infty}^{\infty} p_X(x)\mathrm{d}x = \frac{1}{\sqrt{2\pi}} \int_{-\infty}^{\infty} \exp\left(-\frac{1}{2}u^2\right)\mathrm{d}u.$

Let,

$$I = \int_{-\infty}^{\infty} \exp\left(-\frac{1}{2}u^2\right)\mathrm{d}u \Rightarrow I^2 = \int_{-\infty}^{\infty}\int_{-\infty}^{\infty} \exp\left[-\left(\frac{u^2+v^2}{2}\right)\right]\mathrm{d}u\mathrm{d}v$$

$$u = r\cos\theta, \quad v = r\sin\theta \Rightarrow r\mathrm{d}r\mathrm{d}\theta = \mathrm{d}u\mathrm{d}v$$

$$\Rightarrow I^2 = \int_{-\infty}^{\infty}\int_{0}^{2\pi} r\exp\left(-\frac{r^2}{2}\right)\mathrm{d}r\mathrm{d}\theta = 2\pi$$

3.6 Exponential Random Variable

The exponential random variable is a continuous random variable that has found widespread application in modeling Poisson processes. They are particularly useful for modeling random points in time or space, such as the time at which call requests arrive at a switch, the moment when shooting noise occurs in the photon counting processing of optical devices, and the instant when file requests are made to a service.

The exponential random variable t takes values within the interval $[0, \infty)$, and its continuous distribution function is described by

$$F_T(t) = 1 - \mathrm{e}^{-\lambda t}, \quad t \in [0, \infty) \tag{3.56}$$

where, λ represents the average number of occurrences per unit time, while t is the number of time units before the next event occurs in the modeled process. The PDF for exponential random variables is given as follows:

$$f_T(t) = \lambda\mathrm{e}^{-\lambda t}, \quad t \in [0, \infty) \tag{3.57}$$

The following is Poisson models for counting and exponential models for waiting times. Consider a time interval from 0 to T. Let us randomly place n points within this interval, where $0 < t_1 < t_2 < T$.

Define $\{\text{success}\} = \{\text{point lies in the sub-interval } t_1, t_2\}$

$$P\{success\} = \frac{t_2 - t_1}{T} = p, \quad P\{failure\} = 1 - p$$

Define X as the number of points in t_1, t_2, now consider

$$n \to \infty, \quad \frac{t_2 - t_1}{T} \to 0, \quad \frac{n(t_2 - t_1)}{T} \to a$$

Let $n/T = \lambda, \quad t_2 - t_1 = t_a \Rightarrow a = \lambda t_a$, according to Eq. (3.53) get

$$P(X = k) = e^{-\lambda t_a} \frac{(\lambda t_\alpha)^k}{k!}, \quad k = 0, 1, 2, \cdots \tag{3.58}$$

3.7 Lognormal Random Variable

A lognormal distribution is characterized by a random variable whose logarithm follows a normal distribution. In the short term, the lognormal distribution closely resembles the normal distribution. However, over the long term, it exhibits a greater concentration of values in the upper range.

Some quantities are inherently asymmetrical. For instance, consider the time it takes for individuals to complete a specific task: due to individual differences, we observe a distribution where all values are positive(as time cannot be negative). We can anticipate the likely shape of this distribution: there is a non-reachable minimum time, followed by a few exceptionally fast "champions", then a peak representing the most common completion times for the majority, and finally, a tail of "dropouts" at the end. Clearly, the Gaussian distribution, which allows for both positive and negative values and is symmetric with short tails, does not adequately describe such a scenario.

In many applications, particularly in the fields of reliability and maintainability, data may not adhere to a normal distribution. However, the logarithm of a random variable might, resulting in what is known as a lognormal distribution. When applying a lognormal distribution, plotting the data on lognormal probability paper will yield a straight line. The analysis involves calculating the mean and standard deviation of the logarithmic values and then taking the inverse logarithm of the final outcome.

Given the density function of a lognormal distribution, the expressions for the reliability and unreliability functions of the distribution can be derived from the definitions of these functions.

The lognormal distribution is defined as follows:

Let X be a continuous random variable with a positive value. If In $X \sim N(\mu, \sigma^2)$ follows a normal distribution, then the probability density of X is given by

$$f(x, \mu, \sigma) = \begin{cases} \frac{1}{x\sqrt{2\pi}\sigma} \exp\left[-\frac{1}{2\sigma^2}(\ln x - \mu)^2\right] & x > 0 \\ 0 & x \leq 0 \end{cases} \tag{3.59}$$

It is then said that the random variable X follows a lognormal distribution, denoted as In $X \sim N(\mu, \sigma^2)$.

3.8 Transformation of Random Variables

Suppose there is a functional relationship between two random variables x_1 and x_2, such that $x_1 = g(x_2)$. If the distribution function of x_1 is known to be $f_1(x_1)$, then calculate the distribution function of $(x_2, f_2(x_2))$.

Write two probability distributions in differential form, with

$$D_P = f_1(x_1)\mathrm{d}x_1 \tag{3.60}$$

$$D_P = f_2(x_2)\mathrm{d}x_2 \tag{3.61}$$

If in Eq. (3.61), we replace x_1 with $g(x_2)$ and $\mathrm{d}x_1$ with $g'(x_2)\mathrm{d}x_2$, there is

$$D_P = f_1(g(x_2))g'(x_2)\mathrm{d}x_2 \tag{3.62}$$

By comparing Eq. (3.61) with Eq. (3.62), obtain

$$f_2(x) = f_1(g(x_2))g'(x_2) \tag{3.63}$$

In this manner, we derive the distribution function $f_1(x)$ of x_2.

Let X be a random variable (RV), and define $Y = g(X)$, given the probability density function PDF of X, the PDF of Y is

$$P(y < Y \leq y + \mathrm{d}y) = p_Y(y)\mathrm{d}y$$

$$= p_X(x_1)|\mathrm{d}x_1| + p_X(x_2)|\mathrm{d}x_2| + p_X(x_3)|\mathrm{d}x_3| \tag{3.64}$$

$$\Rightarrow p_Y(y) = \frac{p_X(x_1)}{\left|\frac{dy}{dx}\right|_{x=x_1}} + \frac{p_X(x_2)}{\left|\frac{dy}{dx}\right|_{x=x_2}} + \frac{p_X(x_3)}{\left|\frac{dy}{dx}\right|_{x=x_3}}$$

In general

$$p_Y(y) = \sum_{i=1}^{n} \frac{p_X(x_i)}{\left|\frac{dy}{dx}\right|_{x=x_i}}, \quad x_i = g^{-1}(y) \tag{3.65}$$

Example 3.4

$$Y = \exp(X), \quad X \sim N(m, \sigma)$$

$$-\infty < x < \infty \Rightarrow 0 < y < \infty$$

$$y = \exp(x) \Rightarrow x = \lg y, \quad \frac{dy}{dx} = \exp(x) = y$$

$$p_Y(y) = \left.\frac{p_X(x)}{\exp(x)}\right|_{x=\lg y} = \frac{1}{\sqrt{2\pi}\sigma y}\exp\left[-\frac{1}{2}\left(\frac{\lg y - m}{\sigma}\right)^2\right], \quad 0 < y < \infty$$

Y is supposed to a lognormal distribute random variable.

Example 3.5

$$\text{Let } X \sim N(m, \sigma)$$

$$U = \frac{X - m}{\sigma}$$

Find

$P(m - 3\sigma < X \le m + 3\sigma)$ and $p_U(u)$.

Solution:

From $X \sim N(m, \sigma)$

$$\Rightarrow U = \frac{X - m}{\sigma} \sim N(0, 1)$$

$$\Rightarrow p_U(u) = \frac{1}{\sqrt{2\pi}}\exp\left(-\frac{u^2}{2}\right), \quad -\infty < u < \infty$$

$$\Rightarrow \langle U \rangle = 0, \quad \langle U^2 \rangle = 1$$

$$P_U(u) = \frac{1}{\sqrt{2\pi}}\int_{-\infty}^{u}\exp\left(-\frac{s^2}{2}\right)ds = \frac{1}{\sqrt{2\pi}}\int_{-\infty}^{0}\exp\left(-\frac{s^2}{2}\right)ds + \frac{1}{\sqrt{2\pi}}\int_{0}^{u}\exp\left(-\frac{s^2}{2}\right)ds$$

$$= 0.5 + erfu$$

which

$$erfu = \frac{1}{2\pi}\int_{-\infty}^{\infty} \exp\left(-\frac{s^2}{2}\right)\mathrm{d}s$$

$$\int_{-1}^{1} p_U(u)\mathrm{d}u = 0.682\ 68$$

$$\int_{-2}^{2} p_U(u)\mathrm{d}u = 0.954\ 50$$

$$\int_{-3}^{3} p_U(u)\mathrm{d}u = 0.997\ 30$$

$$\Rightarrow P(m - 3\sigma < X \le m + 3\sigma) = 0.997\ 30$$

$$m \pm 3\sigma \ extremes\ X \sim N(m, \sigma)$$

$$\Rightarrow p_X(x) = \frac{1}{\sqrt{2\pi}\sigma}\exp\left[-\frac{1}{2}\left(\frac{x-m}{\sigma}\right)^2\right], \quad -\infty < x < \infty$$

$$U = \frac{x-m}{\sigma}, \quad -\infty < u < \infty$$

$$x = \sigma u + m$$

$$\frac{\mathrm{d}u}{\mathrm{d}x} = \frac{1}{\sigma}$$

$$p_U(u) = \sigma pX(x = \sigma u + m)$$

$$= \sigma\frac{1}{\sqrt{2\pi}\sigma}\exp\left(-\frac{1}{2}u^2\right) \sim N(0, 1)$$

Example 3.6 Known moment generating function, suppose

$$X \sim N(0, 1)$$

Find: (1)$\langle \exp(sX) \rangle$.

(2) when $X \sim N(\mu, \sigma)$, $\langle \exp(sX) \rangle$.

Solution:

(1)

$$\begin{aligned}
\langle \exp(sX) \rangle &= \frac{1}{\sqrt{2\pi}} \int_{-\infty}^{\infty} \exp(sx) \exp\left(-\frac{x^2}{2}\right) \mathrm{d}x \\
&= \frac{1}{\sqrt{2\pi}} \int_{-\infty}^{\infty} \exp\left(-\frac{x^2 - 2sx}{2}\right) \mathrm{d}x \\
&= \frac{1}{\sqrt{2\pi}} \int_{-\infty}^{\infty} \exp\left[-\frac{(x-s)^2}{2} + \frac{s^2}{2}\right] \mathrm{d}x \\
&= \frac{1}{\sqrt{2\pi}} \int_{-\infty}^{\infty} \exp\left(\frac{s^2}{2}\right) \exp\left[-\frac{(x-s)^2}{2}\right] \mathrm{d}x \\
&= \exp\left(\frac{s^2}{2}\right)
\end{aligned}$$

(2) when $X \sim N(\mu, \sigma)$

$$\langle \exp(sX) \rangle = \frac{1}{\sqrt{2\pi}} \int_{-\infty}^{\infty} \exp(sx) \exp\left[-\frac{1}{2}\left(\frac{x-\mu}{\sigma}\right)^2\right] \mathrm{d}x$$

substitute $\frac{x-\mu}{\sigma} = u$, and proceed

$$\langle \exp(sX) \rangle = \exp\left(s\mu + \frac{s^2\sigma^2}{2}\right)$$

Example 3.7 Markov Inequality.

Let X be a random variable such that it takes non-negative values. That is $P(X < 0) = 0$.

Then, for any $a \geq 0$,

$$P(X \geq a) \leq \frac{\langle X \rangle}{a}$$

Proof:

$$\langle X \rangle = \int_0^\infty x p_X(x)\mathrm{d}x = \int_0^a x p_X(x)\mathrm{d}x + \int_a^\infty x p_X(x)\mathrm{d}x$$

$$\geq \int_a^\infty x p_X(x)\mathrm{d}x \geq \int_a^\infty a p_X(x)\mathrm{d}x = aP(X \leq a)$$

Example 3.8 Chebychev Inequality.

Let X be a random variable with mean value μ and standard deviation σ.

Then

$$P(|X - \mu| \geq k) \leq \frac{\sigma^2}{k^2}$$

Proof:

Consider $(X - \mu)^2$. This is non-negative, by Markov inequality.

$$P(|X - \mu|^2 \geq k^2) \leq \frac{\langle (X - \mu)^2 \rangle}{k^2}$$

$$\left\{(X - \mu)^2 \geq k^2\right\} \Rightarrow |X - \mu| \geq k$$

$$\Rightarrow P(|X - \mu| \geq k) \leq \frac{\sigma^2}{k^2}$$

3.9 Multi-dimensional Random Variables

Consider two random variables X and Y. Define

$$E_1 = \{X \le x\}, \quad E_2 = \{Y \le y\}$$

$$E = E_1 \cap E_2 = \{X \le x \cap Y \le y\}$$

Definition

$$P_{XY}(x, y) = P\{X \le x \cap Y \le y\} = P\{X \le x, Y \le y\} \tag{3.66}$$

Note: comma (,) denotes intersection ($\cap$); $P_{XY}(x, y)$ is joint probability distribution function of X and Y (JPDF).

Definition

$$p_{XY}(x, y) = \frac{\partial^2 P_{XY}(x, y)}{\partial x \partial y} \tag{3.67}$$

Remarks:

(1) Geometric interpretation: place a point randomly in the xy-plane.

$$P_{XY}(x, y) = P(X \le x_1 \cap Y \le y_1)$$

(2) $P_{XY}(x, \infty) = P(X \le x \cap Y \le \infty) = P(X \le x) = P_X(x)$
(3) $P_{XY}(\infty, y) = P(X \le \infty \cap Y \le y) = P(Y \le y) = P_Y(y)$
(4) $P_{XY}(\infty, \infty) = P(X \le \infty \cap Y \le \infty) = P(\Omega) = 1$
(5) $P_{XY}(x, -\infty) = P_{XY}(-\infty, y) = 0$
(6) $P(X \le x_1 \cap Y \le y_2) = ?$

Define

$$s_1 = X \le x_1 \cap Y \le y_2$$

$$s_2 = X \le x_1 \cap Y \le y_1$$

$$s_3 = X \le x_1 \cap y_1 < Y \le y_2$$

$$s_1 = s_2 \cup s_3, \quad s_2 \cap s_3 = \varphi$$

$$P(s_1) = P(s_2) + P(s_3)$$

$$P_{XY}(x_1, y_2) = P_{XY}(x_1, y_1) + P_{XY}(X \le x_1 \cap y_1 < Y \le y_2)$$

$$\Rightarrow P_{XY}(X \le x_1 \cap y_1 < Y \le y_2) = P_{XY}(x_1, y_2) - P_{XY}(x_1, y_1)$$

(7) $P(x_1 < X \le x_2 \cap y_1 < Y \le y_2) = ?$

(8)
$$p_{XY}(x, y) = \frac{\partial^2 P_{XY}(x, y)}{\partial x \partial y} \Rightarrow P_{XY}(x, y) = \int_{-\infty}^{x}\int_{-\infty}^{y} p_{XY}(u, v)\mathrm{d}u\mathrm{d}v$$

$$\Rightarrow P_{XY}(\infty, \infty) = \int_{-\infty}^{\infty}\int_{-\infty}^{\infty} p_{XY}(u, v)\mathrm{d}u\mathrm{d}v = P(\Omega) = 1$$

(9)
$$P_{XY}(\infty, y) = P_Y(y) \Rightarrow P_Y(y) = \int_{-\infty}^{\infty}\int_{-\infty}^{y} p_{XY}(u, v)\mathrm{d}u\mathrm{d}v$$

$$\Rightarrow p_Y(y) = \frac{\mathrm{d}P_Y(y)}{\mathrm{d}y} = \int_{-\infty}^{\infty} p_{XY}(x, y)\mathrm{d}x$$

similarly, $p_X(x) = \int_{-\infty}^{\infty} p_{XY}(x, y)\mathrm{d}y$

$$p_X(x) = \text{MarginalPDFof}X$$
$$p_Y(x) = \text{Marginal PDFof}Y$$

(10) Knowing $p_{XY}(x, y)$, we can find the Marginal PDF. The other way is not true.
(11) $P_{XY}(x, y)$ is monotone nondecreasing in x and y, complete specification of two random variables X and Y through joint probability density function.

3.10 Independence of Random Variables

3.10.1 Definition

Suppose X and Y be two random variables, $F(x, y)$, $F_X(x)$, $F_Y(y)$ are the distribution function and marginal distribution functions of (X, Y), respectively, which means that the random variables X and Y are independent of each other.

Equivalent definition: X and Y are two random variables.If for any real $a < b, \quad c < d$, there has $P\{a < X < b, c < Y < d\} = P\{a < X < b\}P\{c < Y < d\}$, that is, event $\{a < X \le b\}$ is independent of event $\{c < Y \le d\}$, then the random variables X and Y are said to be independent of each other.

3.10.2 Necessary and Sufficient Conditions for Independence

(1) Let (X, Y) be a discrete random variable, and the distribution law is p_{ij}, then

$$X \text{ and } Y \text{ independent of each other} \Leftrightarrow p_{ij} = p_i p_j \tag{3.68}$$

(2) Let (X, Y) be a continuous random variable, with a probability density of $f(x, y)$, then

$$X \text{ and } Y \text{ independent of each other} \Leftrightarrow f(x, y) = f_X(x) f_Y(y) \tag{3.69}$$

From the above conclusion, it can be concluded that to determine the independence of two random variables X and Y, one only needs to determine their respective edge distributions, and then see if the product of the edge distributions for each possible pair of values (X, Y) is equal to the joint distribution.

(3) Independence of n-dimensional Random Variables

Suppose the distribution function of n-dimensional random variables $(X_1, X_2, \cdots, X_n)$ be $F(x_1, x_2, \cdots, x_n)$. If the marginal distribution function of X_k is $F_{X_k}(x_k), \quad k = 1, 2, \cdots, n$, and there is

$$F(x_1, \cdots, x_n) = F_{X_1}(x_1) F_{X_2}(x_2) \cdots F_{X_n}(x_n) \tag{3.70}$$

$X_1, X_2, \cdots, X_n$ are independent of each other, or $(X_1, X_2, \cdots, X_n)$ are independent.

Example 3.9 Let the distribution law of (X, Y) be.

X	Y	
	0	1
0	0.3	0.3
1	0.3	0.1

Find if X and Y are independent.

Solution:

The marginal distribution law of X, Y is:

X	0	1
P	0.6	0.4

Y	0	1
P	0.6	0.4

$$\Rightarrow P(X = 0)P(Y = 0) = 0.6 \times 0.6 = 0.36$$

And

$$P(X = 0, Y = 0) = 0.3$$

$$P(X = 0, Y = 0) \neq P(X = 0)P(Y = 0)$$

X and Y Are not Independent.

Example 3.10 Suppose

$$(X, Y) \sim \begin{cases} 6e^{-(2x+3y)} & x > 0, y > 0 \\ 0 & \text{other} \end{cases}$$

Find if X and Y are independent.

Solution:

When $x \leq 0$, $f_X(x) = 0$.

When $x > 0$, $f_X(x) = \int_0^{+\infty} 6e^{-(2x+3y)}dy = e^{-2x}\left(-2e^{-3y}\right)|_0^{+\infty} = 2e^{-2x}$.

Get $f_X(x) = \begin{cases} 2e^{-2x} & x > 0 \\ 0 & x \leq 0 \end{cases}$

As the same, get $f_Y(y) = \begin{cases} 3e^{-3y} & y > 0 \\ 0 & y \leq 0 \end{cases}$

$$\Rightarrow f(x, y) = f_X(x)f_Y(y)$$

X and Y are independent.

3.11 Joint Distribution Function

The joint distribution function is also known as a multidimensional distribution function. Taking the two-dimensional case as an example, let (X, Y) be a two-dimensional random variable,x, y be any real number, and the binary function: $F(x, y) = P(X \leq x \cap Y \leq y) = P(X \leq x, Y \leq y)$, which is called the distribution function of two-dimensional random variables (X, Y), or the joint distribution function of X and Y.

Consider a function $g(X, Y)$. Definition

$$\langle g(X, Y)\rangle = \int_{-\infty}^{\infty}\int_{-\infty}^{\infty} g(x, y)p_{XY}(x, y)\mathrm{d}x\mathrm{d}y \tag{3.71}$$

Given

(1)

$$m_{nk} = \left\langle X^n Y^k\right\rangle = \int_{-\infty}^{\infty}\int_{-\infty}^{\infty} x^n y^k p_{XY}(x, y)\mathrm{d}x\mathrm{d}y \tag{3.72}$$

Clearly,

$$m_{10} = \langle X\rangle = \int_{-\infty}^{\infty}\int_{-\infty}^{\infty} x p_{XY}(x, y)\mathrm{d}x\mathrm{d}y = \int_{-\infty}^{\infty} x p_X(x)\mathrm{d}x = \eta_x \tag{3.73}$$

Similarly,

$$m_{01} = \eta_y \tag{3.74}$$

(2)

$$\mu_{nk} = \left\langle (X - \eta_x)^n \left(Y - \eta_y\right)^k \right\rangle \tag{3.75}$$

Clearly,

$$\mu_{20} = \sigma_x^2, \quad \mu_{02} = \sigma_y^2 \tag{3.76}$$

Definition

$$\mu_{11} = \left\langle (X - \eta_x)\left(Y - \eta_y\right) \right\rangle = \sigma_{XY} \tag{3.77}$$

σ_{XY} is the covariance of random variables X and Y.
(3) $\gamma_{XY} = \frac{\sigma_{XY}}{\sigma_X \sigma_Y}$, which means correlation coefficient between X and Y.
More on vicariance and correlation coefficient is following.
① Let $Y = aX + b$

$$\langle Y \rangle = a\langle X \rangle + b = a\eta_X + b$$

$$\sigma_Y^2 = \left\langle (Y - \eta_Y)^2 \right\rangle = \left\langle (aX + b - a\eta_X - b)^2 \right\rangle = a^2 \sigma_X^2$$

$$\begin{aligned} \sigma_{XY} &= \langle (X - \eta_X)(Y - \eta_Y) \rangle = \left\langle (X - \eta_X)(aX + b - a\eta_X - b)^2 \right\rangle \\ &= a\langle (X - \eta_X)(X - \eta_X) \rangle = a\sigma_X^2 \end{aligned}$$

$$\gamma_{XY} = \frac{a\sigma_X^2}{\sigma_X \sqrt{a^2 \sigma_X^2}} = \pm 1 (sgna)$$

② $LetX \perp Y \Rightarrow p_{XY}(x, y) = p_X(x) p_Y(y)$

$$\sigma_{XY} = \langle (X - \eta_X)(Y - \eta_Y) \rangle = \int_{-\infty}^{\infty} \int_{-\infty}^{\infty} (x - \eta_X)(y - \eta_Y) p_{XY}(x, y) \mathrm{d}x\mathrm{d}y$$

$$= \int_{-\infty}^{\infty} \int_{-\infty}^{\infty} (x - \eta_X)(y - \eta_Y) p_X(x) p_Y(y) \mathrm{d}x\mathrm{d}y$$

$$= 0 \Rightarrow \gamma_{XY} = 0$$

③ $\sigma_{XY} = \int_{-\infty}^{\infty} \int_{-\infty}^{\infty} (x - \eta_X)(y - \eta_Y) p_{XY}(x, y) \mathrm{d}x\mathrm{d}y$

$$= \int_{-\infty}^{\infty}\int_{-\infty}^{\infty} (x-\eta_X)\sqrt{p_{XY}(x,y)}(y-\eta_Y)\sqrt{p_{XY}(x,y)}\mathrm{d}x\mathrm{d}y$$

$$\leq \left[\int_{-\infty}^{\infty}\int_{-\infty}^{\infty} (x-\eta_X)^2 p_{XY}(x,y)\mathrm{d}x\mathrm{d}y\right]^{\frac{1}{2}}\left[\int_{-\infty}^{\infty}\int_{-\infty}^{\infty} (y-\eta_Y)^2 p_{XY}(x,y)\mathrm{d}x\mathrm{d}y\right]^{\frac{1}{2}}$$

$$= \pm\sigma_X\sigma_Y \Rightarrow -1 \leq \frac{\sigma_{XY}}{\sigma_X\sigma_Y} = \gamma_{XY} \leq 1$$

④ Definition: $\gamma_{XY} = 0 \Rightarrow X$ and Y are uncorrelated; $\gamma_{XY} = \pm 1$ are the limits of linear beavior.

⑤ $X \perp Y \Rightarrow X$ and Y are uncorrelated, X and Y are uncorrelated doesn't mean that $X \perp Y$.

3.12 Function of Random Variables

Definition: let X and Y be two random variables.Define $U = g(X, Y)$ and $V = h(X, Y)$. Given the JPDF of X and Y, what is the JPDF of U and V.

Steps are as follows.

(1) Consider $u = g(x, y)$, $v = h(x, y)$, solve for (x, y) from these equations, Let $(x_i, y_i)_{i=1}^n$ be the roots. Note that n could be ∞.

(2) Determines $J = \begin{pmatrix} \frac{\partial u}{\partial x} & \frac{\partial u}{\partial y} \\ \frac{\partial v}{\partial x} & \frac{\partial v}{\partial y} \end{pmatrix}$ or $J^{-1} = \begin{pmatrix} \frac{\partial x}{\partial u} & \frac{\partial x}{\partial v} \\ \frac{\partial y}{\partial u} & \frac{\partial y}{\partial v} \end{pmatrix}$

(3) Find $p_{UV}(u, v) = \sum_{i=1}^{n} \frac{p_{XY}(x_i, y_i)}{|J|}\Big|_{\substack{x = x_i \\ y = y_i}}$, or $p_{UV}(u, v) = \sum_{i=1}^{n} |J^{-1}| p_{XY}(x_i, y_i)\Big|_{\substack{x = x_i \\ y = y_i}}$

(4) Examine g and h and decide upon limits of u and v.

Note: decide if it is easier to work with $\boldsymbol{J}$ of J^{-1}.

Example 3.11 Known: $U = X + Y$ and $V = X - Y$

$$\begin{pmatrix} X \\ Y \end{pmatrix} \sim N\left(\begin{pmatrix} 0 \\ 0 \end{pmatrix} \begin{pmatrix} \sqrt{2} & 0 \\ 0 & \sqrt{2} \end{pmatrix}\right)$$

Find: $P_{UV}(u, v)$.

Solution:
From

$$u = x + y, \quad v = x - y$$

Get

$$|J| = \begin{vmatrix} 1 & 1 \\ 1 & -1 \end{vmatrix} = 2$$

$$P_{UV}(u, v) = \left.\frac{P_{XY}(x, y)}{|J|}\right|_{\substack{x = \frac{u+v}{2} \\ y = \frac{u-v}{2}}} = \frac{1}{2\pi}\frac{1}{2}\exp\left[-\frac{1}{2}\left(\frac{u+v}{2}\right)^2 - \frac{1}{2}\left(\frac{u-v}{2}\right)^2\right]$$

$$= \frac{1}{4\pi}\exp\left[-\frac{1}{2}\left(u^2 + v^2\right)\right] \quad (-\infty < u < \infty, -\infty < v < \infty)$$

Check

$$\langle U \rangle = \langle X \rangle + \langle Y \rangle = 0$$

$$\langle V \rangle = \langle X \rangle - \langle Y \rangle = 0$$

$$\left\langle U^2 \right\rangle = \left\langle X^2 + Y^2 + 2XY \right\rangle = 2 = \sigma_U^2$$

$$\left\langle V^2 \right\rangle = \left\langle X^2 + Y^2 - 2XY \right\rangle = 2 = \sigma_V^2$$

$$\langle UV \rangle = \left\langle X^2 - Y^2 \right\rangle = 0 = \sigma_{UV}^2$$

$$P_U(u) = \int_{-\infty}^{\infty} p_{UV}(u, v)\mathrm{d}v \sim N(0, \sqrt{2})$$

$$P_V(v) = \int_{-\infty}^{\infty} p_{UV}(u, v)\mathrm{d}u \sim N(0, \sqrt{2})$$

Turn out that U and V are independently, because

$$P_{UV}(u, v) = P_U(u)P_V(v)$$

Example 3.12 Discuss the mean, correlation and covariance functions of the following equation.

$$Z(t) = aX(t) + bY(t)$$

where, a and b are determined.

Discuss:

$$E[Z(t)] = am_X(t) + bm_Y(t)$$

$$R_Z(t_1, t_2) = E\{[aX(t_1) + bY(t_2)][aX(t_2) + bY(t_2)]\}$$

$$= a^2R_X(t_1, t_2) + b^2R_Y(t_1, t_2) + abR_{XY}(t_1, t_2) + baR_{YX}(t_1, t_2)$$

If X and Y are orthogonal, get

$$R_Z(t_1, t_2) = a^2R_X(t_1, t_2) + b^2R_Y(t_1, t_2)$$

If X and Y have no relation, get

$$C_Z(t_1, t_2) = a^2C_X(t_1, t_2) + b^2C_Y(t_1, t_2)$$

Exercise 3

E3.1: Consider a random experiment involving tossing of two dice.Define A is even on die 1, B is even on die 2, C is sum of numbers on die 1and die 2,examine if A, B, and C are independent.

E3.2: Determine the probability distribution function and probability density function of the triangular wave shown in Fig. 3.4. Known A, T, find $P(x)$, $p(x)$.

E3.3: Let the probability density of the random variable X be

$$p(x) = \begin{cases} cx & 0 \le x \le 1 \\ 0 & \text{other} \end{cases}$$

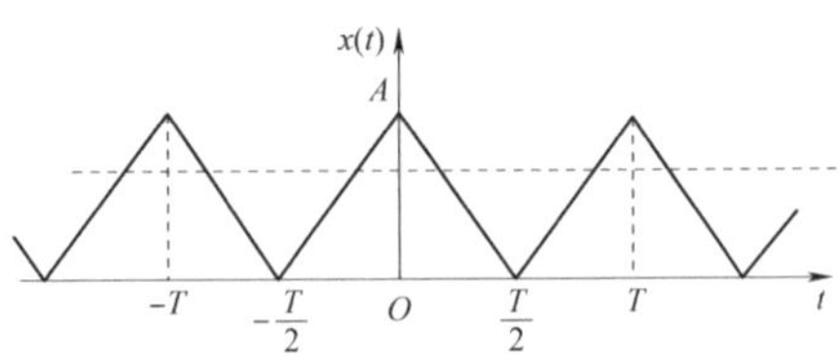

Fig. 3.4 Triangular wave

Try to determine the probability of constants c and X falling within the region (0.3, 0.7).

E3.4: The probability density of the random variable X be

$$p(x) = a\mathrm{e}^{-bx}(a, b \text{ are constant}).$$

(1) Determine the relation between a and b.
(2) When $a = 1$, find the probability distribution function $P(x)$.

E3.5: The probability density function of random variable X is

$$p_X(x) = \begin{cases} \dfrac{A}{\sqrt{1-x^2}} & |x| < 1 \\ 0 & |x| \geq 1 \end{cases}$$

Find A and its probability distribution function.

E3.6: If the probability density function of random variable X is $p_X(x)$, find the probability density function $p_Y(y)$ of $Y = aX + b(k > 0)$.

E3.7: In a random rectification sine function

$$x(\varphi) = A|\sin(\omega t + \varphi)|$$

where, A and ω are constants, φ equals probability random value in $(0, \pi)$. Determine $p(x)$.

Chapter 4
Random Processes

Random processes are mathematical models that encapsulate the randomness observed in the evolution of systems over time. They serve as essential tools for analyzing and predicting the occurrence of random events within these systems. A random process can be viewed as the temporal evolution of a set of random variables or, in other words, as a collection of random events occurring sequentially. Typically, random processes consist of two fundamental components: the state space, which encompasses all possible states of the process, and the transition probability, which quantifies the likelihood of moving from one state to another. Depending on the context of application, stochastic processes can be categorized into discrete-time and continuous-time stochastic processes. Discrete-time processes involve observations at regular time intervals, each associated with a specific state, with the Markov chain being a prime example. In contrast, continuous-time processes allow for observations at any moment, with each instant corresponding to a particular state, exemplified by Brownian motion, where the state space is the set of real numbers and transition probabilities follow a Gaussian distribution.

Random processes find extensive applications across various fields, including finance, physics and engineering. In finance, they are instrumental in modeling the dynamics of stock prices, interest rates, and currency exchange rates. In physics, they describe the erratic motion of particles and the statistical behavior of thermodynamic systems. Engineering applications leverage them to analyze noise, signal integrity, and communication system performance.

J. Zhang and H. Zhou, *Random Vibration and Spectrum Analysis*,
https://doi.org/10.1007/978-981-95-0191-5_4

4.1 Random Process Definition and Statistical Characteristics

1. Random Process Definition

A random process $X(t)$ is a collection of random variables indexed by a real parameter t, often representing time. The set of all potential values that a random process can assume is known as its $\{X(t), t \in T\}$. For instance, the number of customers a store serves between times t_0 and t_k is a set of random variables indexed by time; this ensemble is a random process. It's important to note that the parameter does not have to be time.

A random process is characterized by an interminable family of random variables that are interconnected. When examining random processes, we discern the underlying laws of necessity through apparent randomness, encapsulating these laws in probabilistic terms. The essence of this discipline lies in understanding the necessary from the random.

The foundational theories of random processes were established by Kolmogorov and Dube, with roots in the study of statistical mechanics by Gibbs, Boltzmann, Poincare, and others, as well as the pioneering work on Brownian motion by Einstein, Wiener, Lévy, and their contemporaries.

Studying random processes involves a variety of methods, broadly classified into probability methods, which include orbit properties, stopping times, and stochastic differential equations, and analytical methods, which employ measure theory, differential equations, semi-group theory, functional analysis, and Hilbert spaces. In practice, these methods are often employed in tandem, with combinatorial and algebraic methods also playing significant roles in the study of certain specialized stochastic processes.

2. Description of a Random Process

The description of a random process involves detailing its probability distribution functions, starting with the first order distribution and progressing to higher orders. A complete description of a random process encompasses its probability density functions up to the n-th order.

$$P_X(x, t) = P[X(t) \le x]$$

$$p_X(x, t) = \frac{\partial P_X(x, t)}{\partial x}$$

$$P_{XX}(x_1, x_2; t_1, t_2) = P[X(t_1) \le x_1 \cap X(t_2) \le x_2]$$

$$p_{XX}(x_1, x_2; t_1, t_2) = \frac{\partial^2 P_{XX}(x_1, x_2; t_1, t_2)}{\partial x_1 \partial x_2}$$

$$P_{\tilde{X}}(\tilde{x};\tilde{t}) = P\left[\bigcap_{i=1}^{n}\{X(t_i) \le x_i\}\right]$$

$$p_{\tilde{X}}(\tilde{x};\tilde{t}) = \frac{\partial^n P_{\tilde{X}}(\tilde{x};\tilde{t})}{\partial x_1 \partial x_2 \cdots \partial x_n}$$

Specify $P_{\tilde{X}}(\tilde{x};\tilde{t})$ for all n and for any choice of $\tilde{t}$.
Or
Specify $p_{\tilde{X}}(\tilde{x};\tilde{t})$ for all n and for any choice of $\tilde{t}$.

3. **Statistical Characteristics of Random Processes**

The statistical characteristics of random processes encompass the mean function, variance function, covariance function, and correlation function. These characteristics are defined as follows:

(1) Mean function: the expected value of the process at each point in time.

$$\mu_X(t) = E\{X(t)\}$$

(2) Variance function: a measure of the dispersion of the process around its mean.

$$\sigma_X^2(t) = E\left\{[X(t) - \mu_X(t)]^2\right\}$$

(3) Covariance function: a measure of the linear relationship between the process and a time-shifted version of itself.

$$C_X(s,t) = E\{[X(s) - \mu_X(s)][X(t) - \mu_X(t)]\}$$

(4) Correlation function: a normalized version of the covariance function that reflects the degree of correlation between different points in time.

$$R_X(s,t) = E\{X(s)X(t)\}$$

The relationship between these statistical features is given by the following equations:

$$\begin{aligned} C_X(s,t) &= E\{[X(s) - \mu_X(s)][X(t) - \mu_X(t)]\} \\ &= E\{X(s)X(t)\} - \mu_X(s)\mu_X(t) \\ &= R_X(s,t) - \mu_X(s)\mu_X(t) \\ \sigma_X^2(t) &= R_X(t,t) - [\mu_X(st)]^2 \end{aligned}$$

4. **Classification of Random Processes**

Random processes can be classified based on statistical features and the characteristics of their parameter sets and state spaces. Statistical feature-based classifications include independent incremental processes, Markov processes, second order moment processes, smooth processes, martingales, updating processes, Poisson processes, and Wiener processes. Based on parameter sets and state space features, processes can be divided into discrete and continuous parameter processes, and further into discrete and continuous state space processes. This leads to four primary categories:

(1) Discrete parameter with discrete stochastic processes.
(2) Continuous parameter with discrete stochastic processes.
(3) Discrete parameter with continuous stochastic processes.
(4) Continuous parameter with continuous stochastic processes.

By making various assumptions about the probability structure of a process, a multitude of specialized stochastic processes can be derived.

In addition to the normal and second order processes mentioned, important ones include independent incremental processes, Markov processes, stationary processes, martingale point processes, and branching processes. Two of the most important and fundamental processes that permeate these classes are Brownian motion and the Poisson process. Their structures are relatively simple, making them easy to study and widely applicable. Many other processes can be constructed starting from these two, with Brownian motion characterized by continuous trajectories and the Poisson process by ascending step functions.

4.2 Power Spectrum Density

The quest for a criterion to estimate the severity of vibrations naturally leads to examining characteristics such as the maximum values (primarily the system's acceleration), the mean value of the signal, and the RMS value over an extended period.

If the mean is zero, the RMS value is effectively the standard deviation of the instantaneous signal. Even when two or more signal samples have very different frequency contents, their RMS values can be combined using the square root of the sum of their squares.

Consider a signal $x(t)$ and apply a series of rectangular filters with central frequencies f and bandwidths Δf, as depicted in Fig. 4.1.

The curve that would yield the RMS value with respect to X_{RMS} would provide a description of $x(t)$'s spectrum.The outcome would vary depending on Δf.

To circumvent this issue, consider the variation of $\frac{X_{\text{RMS}}^2}{\Delta f}$ with f. The RMS value is squared, analogous to power.

4.2.1 *Definition*

Figure 4.2 illustrates the circuit with voltage $u(t)$ applied across resistance R, producing current $i(t)$. The energy dissipated in the resistance over time $\mathrm{d}t$ is given

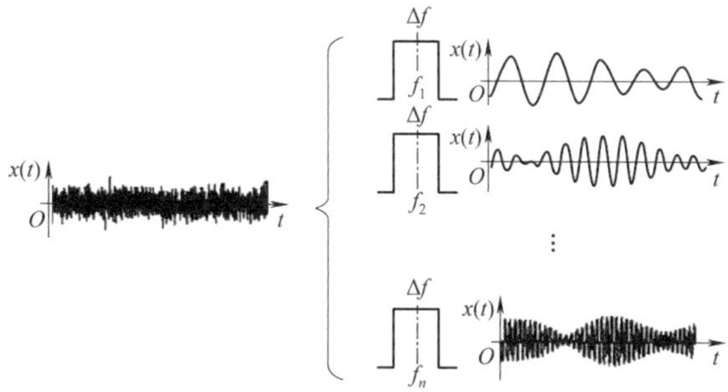

Fig. 4.1 Filter of the random signal

by

$$dE = Ri^2(t)dt = i^2(t) \tag{4.1}$$

The instantaneous power is

$$P(t) = \frac{dE}{dt} = i^2(t)$$

and the energy dissipated between t and $t + T$ is

$$E_T = \int_t^{t+T} i^2(t)dt$$

The mean power in the interval T is

$$P_{m_T} = \frac{1}{T}\int_t^{t+T} p(t)dt = \frac{1}{T}E_T$$

The total energy E and total mean power P_m are

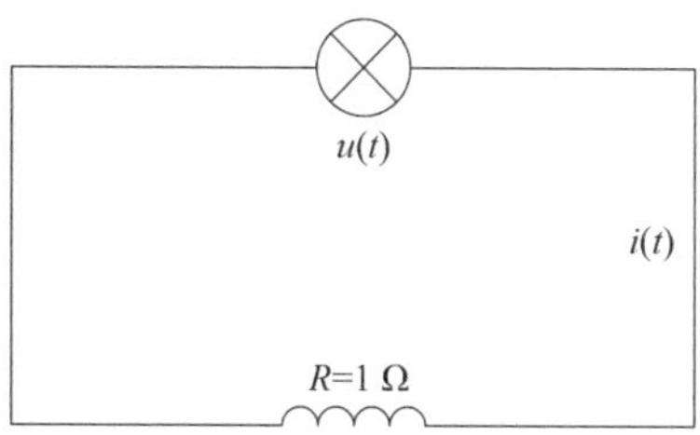

Fig. 4.2 Circuit with voltage and resisance

$$E = \int_{-\infty}^{+\infty} i^2(t)\mathrm{d}t$$

$$P_\mathrm{m} = \lim_{T\to\infty} \frac{1}{T} \int_{-T/2}^{+T/2} i^2(t)\mathrm{d}t$$

The mean power of the excitation $f(t)$ between $-T/2$ and $+T/2$ is

$$P_\mathrm{m} = \lim_{T\to\infty} \frac{1}{T} \int_{-T/2}^{+T/2} |f_T(t)|^2\mathrm{d}t$$

where

$$\begin{cases} f_T = f(t) & |t| \le T/2 \\ f_T = 0 & |t| > T/2 \end{cases}$$

Suppose $f_T(t)$ has a Fourier transform $f_T(f)$. According to Parseval's equality:

$$\int_{-\infty}^{+\infty} |f_T|^2\mathrm{d}t = \int_{-\infty}^{+\infty} |F_T(f)|^2\mathrm{d}f$$

Since when $T \to \infty$,

$$\int_{-T/2}^{+T/2} |f_T|^2\mathrm{d}t = \int_{-\infty}^{+\infty} |f_T(t)|^2\mathrm{d}t$$

$$P_\mathrm{m} = \lim_{T\to\infty} \frac{1}{T} \int_{-\infty}^{+\infty} |F_T(f)|^2\mathrm{d}f = \lim_{T\to\infty} \frac{2}{T} \int_{0}^{\infty} |F_T(f)|^2\mathrm{d}f \tag{4.2}$$

This relation reveals the mean power contained in $f(t)$ when all frequencies are onsidered. If the excitation $f(t)$ is applied to a linear system with constant parameters, whose weighting function is $h(t)$ and the system function is $H(t)$, the response $x_T(t)$ is given by the convolution integral:

$$x_T(t) = \int_{0}^{\infty} h(\tau)f_T(t-\tau)\mathrm{d}\tau \tag{4.3}$$

The mean power of the response is expressed as

$$P_{\text{m,response}} = \lim_{T\to\infty} \frac{2}{T} \int\limits_{0}^{T} x_T^2(t)\mathrm{d}t \tag{4.4}$$

According to Parseval's equality,

$$P_{\text{m,response}} = \lim_{T\to\infty} \frac{2}{T} \int\limits_{0}^{T} |R_T(f)|^2\mathrm{d}f \tag{4.5}$$

where

$$R_T(f) = H(f)F_T(f)$$

We define

$$G_T(f) = \frac{2|F_T(f)|^2}{T} \tag{4.6}$$

The mean power corresponding to the record $f_T(t)$, of finite length T, centered on f, is

$$P_T(f, \Delta f) = \int\limits_{f-\Delta/2}^{f+\Delta/2} G_T(f)\mathrm{d}f \tag{4.7}$$

And the total mean power

$$P_{\text{m}}(f, \Delta f) = \lim_{T\to\infty} \int\limits_{f-\Delta f/2}^{f+\Delta f/2} G_T(f)\mathrm{d}f \tag{4.8}$$

One terms the quantity of the power spectral density (PSD).

$$G(f) = \lim_{t\to\infty} \frac{P_{\text{m}}(f, \Delta f)}{\Delta f} \tag{4.9}$$

An example of track irregularity PSD is provided in Fig. 4.3.

The PSD of $G(f)$ can be written as

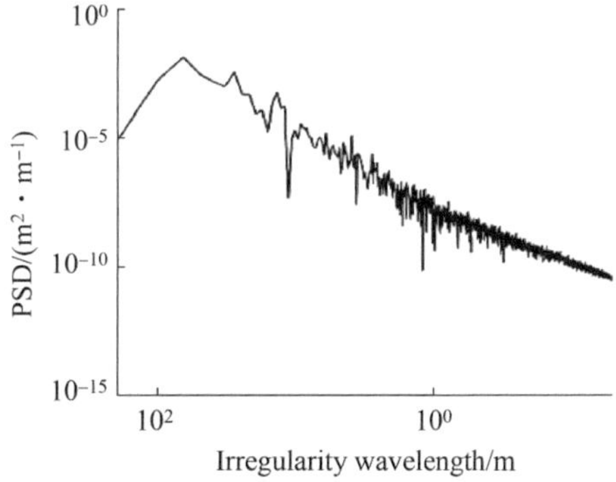

Fig. 4.3 Example of track irregularity PSD

$$G(f) = \lim_{\substack{\Delta f \to 0 \\ T \to \infty}} \frac{1}{T \Delta f} \int_0^T f_T^2(t, \Delta f) \mathrm{d}t \tag{4.10}$$

The signal's segment that lies within the specified frequency range is of particular interest.

Equation (4.10) theoretically defines the PSD. However, in practice, achieving this ideal is unfeasible, as calculating the PSD necessitates an infinite duration of integration and an infinitely narrow bandwidth.

The utility of the auto-power spectrum is extensive, encompassing:

(1) Analyzing the components and structure of vibration frequencies, such as assessing the stability and smoothness of track vehicles, engine resonance frequencies, etc.
(2) The auto-power spectrum also elucidates the vibrational energy and amplitude of loads across various frequency components, thereby facilitating the determination of the load spectrum. This is essential for designing component strength, studying structural responses to external stimulus, and ensuring their longevity.
(3) Fault diagnosis and analysis. For instance, critical equipment like steam turbines, airplanes, high-speed trains, and other rail transit vehicles, as well as machine tools or significant components (e.g., gearboxes), can identify the cause of malfunctions by observing changes in the auto spectrum (the presence of additional spectral peaks).
(4) In the medical field, auto spectrum analysis can be applied to brain wave and Electrocardiogram (ECG) signals to investigate symptoms and underlying pathologies.
(5) Militarily, spectrograms can assist in detecting and identifying the types of submarines.

4.2.2 Some Common PSD Plots

1. **Broadband Process**

A broadband process is a stochastic stationary process with a power spectral density $S(\Omega)$ that maintains a significant value across a frequency band or domain that is strictly the same order of magnitude as the band's center frequency. An example of such a process is the pressure fluctuations on the skin of a missile rocket (jet noise and turbulence of the supersonic boundary layer), as shown in Fig. 4.4.

2. **White Noise**

When analyzing, it is common to idealize the broadband process by assuming a uniform spectral density. For wide-sense stationary signals, there is always a process (illustrated in Fig. 4.5):

$$R(\tau) = \frac{N_0}{2}\delta(\tau), \quad \text{or} \quad S(\omega) = \frac{N_0}{2}$$

It is referred to as the stationary white noise signal, or simply as white noise or white signal. Non-white noise signals are termed colored noise.

White noise is occasionally described as "pure randomness" due to the following attributes:

(1) It is an ideal random signal with an infinite bandwidth.
(2) Its power (e. g., variance) is infinite.
(3) It exhibits no correlation across different time points.

If each random variable of the white noise adheres to a Gaussian distribution, it is termed Gaussian White Noise (WGN). WGN is also an independent signal, representing the epitome of signal randomness.

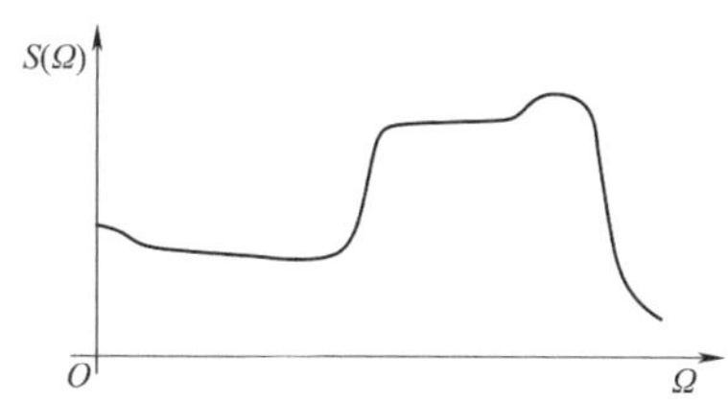

Fig. 4.4 The broadband process

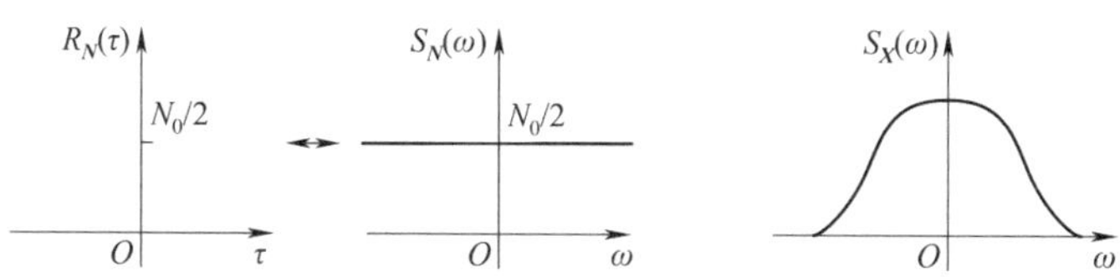

Fig. 4.5 White noise

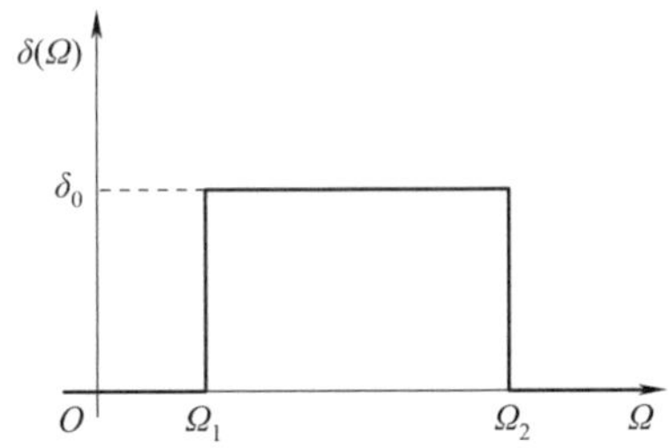

Fig. 4.6 The band-limited white noise

A sequence that consistently maintains independence, $R_m = \frac{N_0}{2}\delta_m$, or $S(e^{i\omega}) = \sum_m R_m e^{-i\omega m} = N_0/2$, whether it is a sequence of random variables or a sequence of values, is known as a Gaussian white noise sequence. The independence of this sequence simplifies the derivation of its probability density functions.

3. **Band-Limited White Noise**

In practical applications, band-limited white noises, such as that in Fig. 4.6, are often used in calculations, providing accurate approximations for many random processes on exciters.

4. **Narrow-Band Process**

A narrow-band process (Fig. 4.7) is a random stationary process with a PSD that has significant values in a single frequency band or a frequency domain whose width is small compared to the value of the band's center frequency.

The signal as a function of time $x(t)$ resembles a sinusoid with an angular frequency ω_0, featuring a randomly varying amplitude and phase. There is typically a single peak between each pair of zero crossings.

By examining individual cycles and envelopes, as seen in Fig. 4.8, these processes are particularly relevant to the response of lightly damped mechanical systems when subjected to broad-band noise excitation.

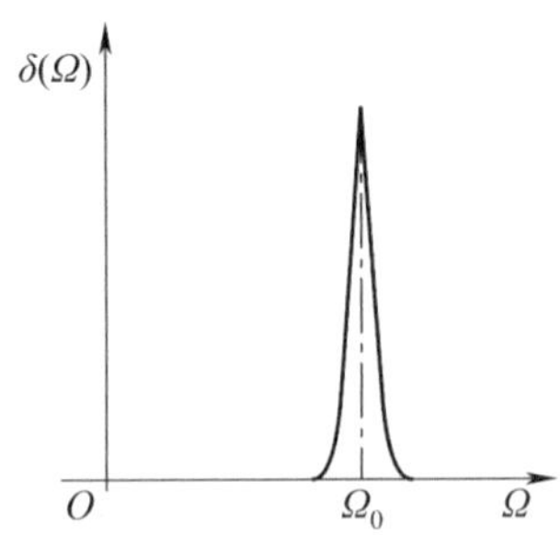

Fig. 4.7 PSD of narrow-band noise

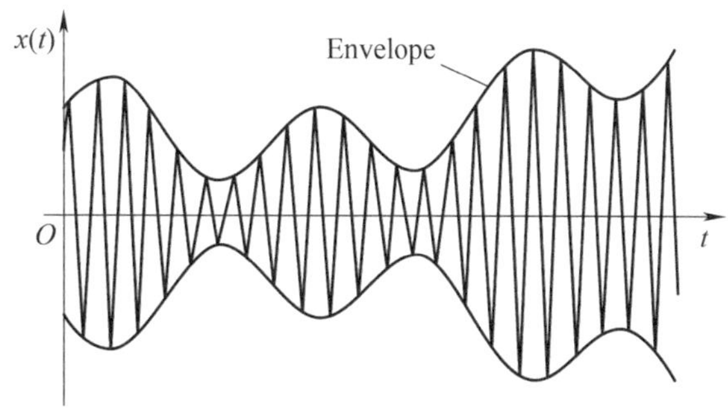

Fig. 4.8 The narrow-band noise

4.2.3 *Computation of PSD*

The analysis of random vibrations typically assumes the process is stationary and ergodic. This assumption allows for the use of a single sample with a finite duration T to represent the statistical characteristics of a larger ensemble of signals. There are several methods to calculate the PSD of this sample.

To analyze a signal, sampling involves converting a continuous vibratory signal into a series of regularly spaced sample points in time. To accurately represent the digitized signal, the sampling frequency must be sufficiently high relative to the highest frequency of the signal under analysis. The sampling frequency must satisfy the sampling theorem, which states that the lowest sampling frequency must be at least twice the highest frequency of the sampled signal.

1. **Calculation of PSD from RMS Value of Filtered Signal**

The theoretical Eq. (4.10), which assumes an infinite duration and a zero analysis bandwidth, is replaced by the approximate relation:

$$G(f) = \frac{1}{T\Delta f}\int_0^T x_{\Delta f}^2(t,f)\mathrm{d}t = \frac{\overline{x_{\Delta f}^2}}{\Delta f} \tag{4.11}$$

where $x_{\Delta f}^2$ is the mean square value of the samples of T, calculated at the output of a filter of central frequency f and non-zero width Δf.

2. **Calculation of PSD from Fourier Transform**

By definition, the PSD of the random process is, over time $T, G(f) = \sum_{i=1}^{n} G_T^i(f)/n$, n being the number of functions $x^i(t)$, and for infinite $T, G(f) = \lim_{T\to\infty} G_T(f)$. If the process is stationary and ergodic, the PSD can be calculated from several samples of a single recording.

$$G_{ff}(f) = \lim_{T\to\infty}\frac{2}{T}E\left[|X(f,T)|^2\right] \tag{4.12}$$

The calculation data typically encompass the following elements:

① The upper limit of the frequency spectrum;
② The quantity of data points in PSD, or equivalently, the frequency resolution;
③ The maximum allowable statistical error.

(1) Upper Limit of the Frequency Spectrum

Given a sampled signal with a frequency of f_{samp} and adhering to the principles of the sampling theorem, the PSD is applicable solely to signals with frequencies less than $f_{\max} = f_{\text{samp}}/2.6$. The factor of 2.6 is derived from the Shannon sampling theorem, which represents the minimum requirement for the theorem to hold.

(2) Sample Extraction Over Duration T

There are two prevailing approaches to calculating the PSD: one postulate that the signal is periodic, consisting of repeated segments each lasting T seconds; the other postulate that the signal's value is zero at all times outside the interval of the sample under consideration.
These methodologies are deemed equivalent. In both scenarios, signal isolation is effected by truncating a segment of the signal, a process akin to applying a rectangular window function with a uniform amplitude of 1 over the duration of the sample and zero elsewhere.

If $f(t)$ represents the signal to be analyzed, the Fourier transform is then computed using $x(t) = f(t)r(t)$.

3. **Averaging**

To achieve the best possible resolution with the data available in our calculations, we aim to plot the PSD with the smallest possible frequency step. For samples with a duration of T the step cannot be less than $1/T$. The accuracy obtained with this resolution is often unacceptable. Several solutions are available:

(1) Measure the phenomenon multiple times, calculate the duration of each PSD sample as T, and average the obtained spectra.
(2) If only one sample of duration T is available, limit the resolution by accepting a larger analysis step Δf and averaging it.
(3) Alternatively, by calculating the average values of several closely related frequency components, when the noise under analysis can be compared with white noise, the considered spectral components are separated by an interval of $1/T$. If averaging is performed on K PSD, the obtained average will be allocated to the center frequency of the interval with a width equal to K/T, which characterizes the effective resolution of the PSD calculated from this.

By dividing the initial sample of duration into sub-samples (or blocks), these will be utilized to calculate spectra of a specified resolution and their average, $\frac{1}{K}\sum_i G_i(f)$.

The results are the same for a given duration and a given resolution. This is the most commonly used final step. The window (whether rectangular or not) is applied to each block.

4.2.4 Variance of Measured Power Spectral Density

The variance of $\hat{G}(f)$ is given by:

$$s^2_{\hat{G}(f)} = var\left[\frac{G(f)\chi^2_n}{n}\right]$$

$$s^2_{\hat{G}(f)} = \left[\frac{G(f)}{n}\right]^2 var\left(\chi^2_n\right) \tag{4.13}$$

However, the variance of a chi-square distribution is equal to twice the number of degrees of freedom:

$$var\left(\chi^2_n\right) = 2n \tag{4.14}$$

namely,

$$s^2_{\hat{G}(f)} = 2\left[\frac{G(f)}{n}\right]^2 n = 2\frac{G^2(f)}{n} \tag{4.15}$$

It means that this law is equal to n.

$$\overline{\hat{G}(f)} = \frac{G(f)}{n}\overline{\chi^2_n} = \frac{G(f)}{n}n$$

$$\hat{G}(f) = G(f) \tag{4.16}$$

The statistical error is

$$\varepsilon^2 = \frac{s^2_{G(f)}}{G(f)} = \frac{2}{n}$$

$$\varepsilon = \sqrt{\frac{2}{n}} \tag{4.17}$$

ε is called "standard error".

The following will introduce the relationship between number of degrees of freedom, duration and bandwidth of analysis. The relation can be obtained by $E\left\{[\hat{G}(f) - G(f)]^2\right\}$, or from the autocorrelation function.

(1) By $E\left\{[\hat{G}(f) - G(f)]^2\right\}$

We know that

$$E\left\{[\hat{G}(f) - G(f)]^2\right\} \approx \frac{G^2(f)}{T\Delta f} + \frac{\Delta f^2}{576}\left[G^*(f)\right]^2 \tag{4.18}$$

The first item is variability and the second item is bias at the right end of Eq. (4.18). Except when the slope of PSD varies greatly with Δf, the deviation is usually negligible. So

$$\varepsilon^2 = \frac{E\{[\hat{G}(f) - G(f)]^2\}}{G^2(f)} \approx \frac{1}{T\Delta f}$$

This relation is a good approximation as long as ε is lower than approximately 0.2.

(2) From the Autocorrelation Function

$x(t)$ is a vibratory signal response collected at the output of a filter of width Δf. The mean square value of $x(t)$ is given by

$$\lambda_{\Delta f}^2 = \frac{1}{T}\int_0^T x^2(t)\mathrm{d}t$$

Setting $\overline{x_{\Delta f}^2}$ the measured value of $\lambda_{\Delta f}^2$, from definition, we have

$$\begin{aligned}
\varepsilon &= \frac{\overline{\left[\left(\lambda_{\Delta f}^2 - \overline{x_{\Delta f}^2}\right)^2\right]}^{1/2}}{x_{\Delta f}^2} \\
&= \frac{\left[\overline{\lambda_{\Delta f}^4} - 2\lambda_{\Delta f}^2\overline{x_{\Delta f}^2} + \left(\overline{x_{\Delta f}^2}\right)^2\right]^{1/2}}{\overline{x_{\Delta f}^2}} \\
&= \frac{\left[\overline{\lambda_{\Delta f}^4} - \left(\overline{x_{\Delta f}^2}\right)^2\right]^{1/2}}{\overline{x_{\Delta f}^2}}
\end{aligned} \tag{4.19}$$

Equation (4.19) can be written as

$$\begin{aligned}
\overline{\lambda_{\Delta f}^4} - \left(\overline{x_{\Delta f}^2}\right)^2 &= \frac{1}{T}\int_0^T x^2(u)\mathrm{d}u \cdot \frac{1}{T}\int_0^T x^2(v)\mathrm{d}v - \left(\overline{x_{\Delta f}^2}\right)^2 \\
&= \frac{1}{T^2}\int_0^T \mathrm{d}u \cdot \frac{1}{T}\int_0^T \left[\overline{x^2(u)x^2(v)} - \left(\overline{x_{\Delta f}^2}\right)^2\right]\mathrm{d}v
\end{aligned}$$

When set $t = u$ and $\tau = v - u = v - t$,

$$\overline{\lambda_{\Delta f}^4} - \left(\overline{x_{\Delta f}^2}\right)^2 = \frac{1}{T^2}\int_0^T \mathrm{d}t \cdot \frac{1}{T}\int_{-t}^{T-t}\left[\overline{x^2(t)x^2(t+\tau)} - \left(\overline{x_{\Delta f}^2}\right)^2\right]\mathrm{d}\tau$$

namely,

$$\overline{\lambda_{\Delta f}^4} - \left(\overline{x_{\Delta f}^2}\right)^2 = \frac{2}{T^2}\int_0^T \mathrm{d}t \times \frac{1}{T}\int_{-t}^{T-t}\left(\overline{x_{\Delta f}^2}\right)^2 \rho^2(\tau)\mathrm{d}\tau \tag{4.20}$$

where $\rho(\tau)$ is the auto-correlation coefficient. To a narrow band random signal, the coefficient $\rho(\tau)$ is symmetrical when $\tau = 0$ and that $\rho(\tau)$ decrease when $|\tau|$ becomes larger. If T is enough large, the majority of the values of t :

$$\varepsilon^2 = \frac{\overline{\lambda_{\Delta f}^4} - \left(\overline{x_{\Delta f}^2}\right)^2}{\left(\overline{x_{\Delta f}^2}\right)^2} = \frac{2}{T^2}\int_0^T \mathrm{d}t \int_{-\infty}^{\infty} \rho^2(\tau)\mathrm{d}\tau \tag{4.21}$$

where ε is also called standardized variance.

$$\varepsilon^2 = \frac{2}{T}\int_{-\infty}^{\infty} \rho^2(\tau)\mathrm{d}\tau = \frac{4}{T}\int_0^{\infty} \rho^2(\tau)\mathrm{d}\tau$$

$$\varepsilon = 2\sqrt{\frac{1}{T}\int_{-\infty}^{\infty} \rho^2(\tau)\mathrm{d}\tau} \tag{4.22}$$

4.2.5 Generation of Random Signal of Given PSD

1. **Method**

The method for generating random signals with a given duration T from a PSD with maximum frequency $f_{\max}$ includes the following steps:

(1) Computation of the time step $\delta t = 1/f_{\text{samp}} = 1/(2.6 f_{\text{samp}})$.
(2) Selection of the number of PSD defined points M (power of 2).
(3) Calculation of the number of signal points $N = T/(\delta t)$.
(4) Normally, modify N (and duration) and/or M to consider the maximum statistical error ε_0(for future PSD calculation of generated signals), from the relationship $M/N \leq \varepsilon^2/2$, and keep M equal to the power of 2.

(5) Compute the frequency interval between two points of the PSD, $\Delta f = f_{\text{samp}}/(2M)$.
(6) For each of the M points of the PSD, compute a "sinusoid" at $t = k\delta t$.

$$x^m(t) = x^m_{\max} \sin(2\pi ft + \varphi_m), \quad f_m = m\mathrm{d}f\,(1 \le m \le M)$$

(7) Amplitude $\sqrt{2G(f_m)\Delta f}$ [$G(f_m)$ is the value of the given PSD at f_m, the amplitude of a sinusoid being equal to twice its RMS].
(8) Random phase φ_m is a function of the specified distribution law for the instantaneous values of the signal.
(9) Sum of the sinusoids at each time.

2. **Random Phase**

(1) Gaussian Law

The research results indicate that a normal distribution of instantaneous signal values can be obtained when the phase is equal to,

$$\varphi_m = 2\pi\sqrt{-2\ln\tau_1}\cos(2\pi\tau_2) \tag{4.23}$$

or

$$\varphi_m = 2\pi\sqrt{-2\ln\tau_1}\sin(2\pi\tau_2) \tag{4.24}$$

In Eqs. (4.23) and (4.24), τ_1 and τ_2 are two random numbers obeying a rectangular distribution in the interval.

(2) Other Laws

A signal is created here, which instantaneous value follows a given distribution law $F(x)$. This function is non decreasing, where $x \le X$ is equal to

$$P(x \le X) = P[F(x) \le F(X)] \tag{4.25}$$

Set $F(x) = r$, where r is a random variable uniformly distributed with respect to [0, 1]. Then it becomes

$$P[F(x) \le F(X)] = P[r \le F(X)] \tag{4.26}$$

According to the definition of uniform distribution, $P(r < R) = R$, where R is any number between 0 and 1, resulting in $P(x < X) = P[r < F(x)] = F(x)$. In order to generate signals with a distribution of $F(x)$, it is necessary to

$$F(x) = r \tag{4.27}$$

Certainly, to solve the problem by

$$F(x) = 1 - r \tag{4.28}$$

4.3 Ergodicity of a Random Process

The concept of random processes implies the consideration of a large set of sample functions. To determine the statistical characteristics of random processes, it is necessary to observe a multitude of sample functions. Mathematical expectations, variances, correlation functions, and so on are numerical features obtained by averaging the values of a large number of sample functions at a specific time using statistical methods. This average is referred to as a statistical average or ensemble average. Clearly, taking statistical averages requires significant experimental work and complex processing methods. This naturally leads to the question of whether a simpler method can replace the aforementioned method based on the fact that the statistical characteristics of stationary random processes are independent of the starting point of time recording.

Wiener-Khitchine theorem proves that under certain conditions, any sample function of a stationary random process, when observed over a sufficiently long period, converges to the statistical average of the process in a probabilistic sense. Such a random process is termed ergodic.

The ergodicity of each state in a random process can be understood as follows: all sample functions of the random process have equally experienced the various possible states of the process. Therefore, all statistical information of a random process can be derived from any single sample function, and the characteristics of any sample function can fully represent the characteristics of the entire random process.

To facilitate the analysis of the statistical properties of random processes based on sample functions, the concept of ergodicity is introduced. In other words, on the subscript set T, a random process traverses all its possible states according to its distribution function. For ergodic stochastic processes, the set average of the process is equal to the average of any sample on the actual T.

4.3.1 Definitions of Ergodicity

1. **Definitions**

Let $x(t)$ be a sample realization of the random process $X(t)$. We define the time average of a given function of $X(t)$, $g[X(t)]$, as follows:

$$T_{\text{av}}\{g[X(t)]\} = \frac{1}{T}\int_{0}^{T} g[x(t)]\mathrm{d}t$$

If $X(t)$ is an ergodic random process, then $\langle g[X(t)]\rangle = T_{\text{av}}\{g[X(t)]\}$.

There are two definitions of ergodicity.

(1) Definition of Strict (Narrow) Ergodicity

If a random process $X(t)$, whose various time averages (long enough) converge to the corresponding set averages with probability 1, then it is said to have strict ergodicity and is called a strict ergodic process.

(2) Definition of Broad (Generalized) Ergodicity

Let $X(t)$, be a stationary random process, and if its mean and correlation functions have ergodicity, then it is called a wide ergodic process, or simply an ergodic process.

Some points worth noting:

① The above list of definitions of ergodicity is not exhaustive; several other similar definitions can be constructed by considering other descriptors of the random process.
② Ergodic processes are necessarily stationary in nature; a stationary random process need not be ergodic.
③ Physically, ergodicity means that a sufficiently long record of a stationary random process contains all the statistical information about the random phenomenon.

2. **Ergodicity in Mean**

Defined $\overline{X(t)} = \lim\limits_{T\to\infty} \frac{1}{2T}\int_{-T}^{T} X(t)\mathrm{d}t$ as the time average of a stochastic process, it is considered ergodic in mean. If it converges to the ensemble mean with probability 1, that is

$$A(X(t)) = \overline{X(t)} = E[X(t)] = m_X$$

$$T_{\text{av}}\{X(t)\} = \frac{1}{T}\int_{0}^{T} x(t)\mathrm{d}t = \langle X(t)\rangle$$

A stationary process is deemed ergodic in mean, when the mean value encapsulated within any sample function reflects the same as all states of the stochastic process (irrespective of time), and the frequency of occurrence aligns with the probability of each state within the process.

3. **Ergodicity in Autocorrelation**

Defined

$$R_X(t, t+\tau) = \overline{X(t)X(t+\tau)} = \lim_{T\to\infty} \frac{1}{2T} \int_{-T}^{T} X(t)X(t+\tau)\mathrm{d}t$$

as the Time Autocorrelation Function (TAF) of Stochastic Processes, it is said to possess ergodicity if it converges to the ensemble mean with probability 1.

$$R_X(t, t+\tau) = \overline{X(t)X(t+\tau)} = E[X(t)X(t+\tau)] = R_X(\tau)$$

For a stationary random process, if its autocorrelation function exhibits ergodicity, then the mean square value of the process is also deemed ergodic.

$$R_X(\tau) = \lim_{T\to\infty} \frac{1}{2T} \int_{-T}^{T} x(t+\tau)x(t)\mathrm{d}t$$

Every sample function of such a process will have experienced the full spectrum of second- order possible states.

Figure 4.9 provides a schematic representation distinguishing between ergodic and non-ergodic processes in terms of state occupancy.

While the time average of a general stochastic process is a random variable, the time average of an ergodic process is a deterministic quantity. Thus, the time average of any single sample path can be utilized as a proxy for the statistical mean of the entire stochastic process. In practical scenarios, although the time horizon T is finite and cannot extend to infinity, it should be sufficiently long.

In practice, proving the ergodicity of each state of a stationary random process is challenging. However, based on empirical evidence, we often operate under the assumption that the process is ergodic across all states. The attributes of any given

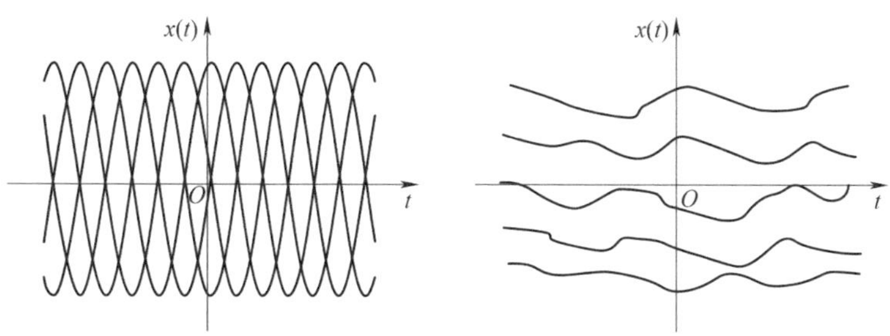

(a) Ergodic process (b) Non-ergodic process

Fig. 4.9 Schematic diagram of ergodic and non-ergodic processes

sample function are assumed to encapsulate the full spectrum of statistical characteristics of the stochastic process, thereby simplifying research procedures and practical statistical methodologies. It is a common belief in most practical systems that noise and signals are generally stationary and ergodic.

4. **Criterion Theorem for Ergodic Processes (Ergodicity of Each State)**

(1) Discriminant Theorem of Mean Ergodicity

A necessary and sufficient condition for the mean function of a stationary random process to be ergodic is

$$\lim_{T\to\infty}\frac{1}{T}\int_{0}^{2T}\left(1-\frac{|\tau|}{2T}\right)\left[R_X(\tau)-m_X^2\right]\mathrm{d}\tau=0$$

(2) Discriminant Theorem of Autocorrelation Ergodicity

A necessary and sufficient condition for the autocorrelation function of a stationary random process to be ergodic is

$$\lim_{T\to\infty}\frac{1}{T}\int_{0}^{2T}\left(1-\frac{\tau_1}{2T}\right)\left[B(\tau_1)-R_X^2(\tau)\right]\mathrm{d}\tau_1=0$$

where, $B(\tau_1)=E[X(t+\tau+\tau_1)X(t+\tau_1)X(t+\tau)X(t)]$.

For a normally distributed stationary random process, if the mean is zero and the autocorrelation function $R_X(\tau)$ is continuous, it can be proven that a sufficient condition for the process to be ergodic is

$$\int_{0}^{\infty}|R_X(\tau)|\mathrm{d}\tau<\infty$$

4.3.2 PSD and Cross-PSD

1. **PSD of a Random Process**

Consider the case where the function of time belongs to a random process, where each function will be noted $rx(t)$. A sample of this signal of duration T, will be denoted by $nx_T(t)$, and its Fourier transform $rX_T(\omega)$. Its PSD is

$$rS_T(\omega) = \frac{2|rX_T(\omega)|^2}{T}$$

The PSD of the random process over time T is

$$S_T(\omega) = \frac{\sum_{i=1}^{n} rX_T(\omega)}{n}$$

where, n is the number of functions $rx(t)$ and, for T infinite,

$$S(\omega) = \lim_{T\to\infty} S_T(\omega) \tag{4.29}$$

If the process is stationary and ergodic, the PSD of the process can be calculated starting from several samples of one recording only.

2. **Cross-PSD**

From two samples of random signal records $x(t)$ and $y(t)$, the cross-power spectrum is defined by

$$S_{XY}(\omega) = \lim_{T\to\infty} \frac{2X(\omega)Y(\omega)}{T} \tag{4.30}$$

If the limit exists, and $X(\omega)$ and $Y(\omega)$ being the Fourier transforms of $x(t)$ and $y(t)$ calculated between 0 and T, respectively.

The cross-power spectrum of stationary signals is also found by the FFT of the correlation function.

$$\begin{cases} S_{XY}(\omega) = \int\limits_{-\infty}^{+\infty} R_{XY}(\tau)\mathrm{e}^{-i\omega\tau}\mathrm{d}\tau \\ S_{YX}(\omega) = \int\limits_{-\infty}^{+\infty} R_{YX}(\tau)\mathrm{e}^{-i\omega\tau}\mathrm{d}\tau \end{cases} \tag{4.31}$$

The physical senses of the cross-power spectrum are:

(1) If $S_{XY}(\omega)$ is very large, it shows that the correlation degree of the corresponding frequency components is very high.
(2) If $S_{XY}(\omega)$ is zero, it shows that the correlation degree of the corresponding frequency components is orthogonal.

The cross-power spectrum is symmetric

$$S^*_{XY}(\omega) = S_{YX}(\omega), \quad S^*_{XY}(\omega) = S_{XY}(-\omega) \tag{4.32}$$

From Eq. (4.32): the real part of the two cross-power spectra is the same but the sign of the imaginary part is opposite; and the correlation function of real signals is a real function, so the real part of the cross-power spectrum is an even function, and the imaginary part is an odd function.

3. **Relation between PSD and Correlation Function**

For a stationary process, its PSD is

$$S(\omega) = \frac{1}{2\pi} \int_{-\infty}^{+\infty} R(\tau) e^{-i\omega\tau} d\tau \tag{4.33}$$

where $R(\tau)$ being an even function of τ. Taking the inverse transform of $S(\omega)$ in Eq. (4.34):

$$R(\tau) = \int_{-\infty}^{+\infty} S(\omega) e^{-i\omega\tau} d\omega \tag{4.34}$$

Since $S(\omega)$ is an even function, get

$$R(\tau) = 2 \int_{0}^{+\infty} S(\omega)\cos(\omega\tau) d\omega$$

When $\tau = 0$

$$R(0) = \overline{x^2(t)} = 2 \int_{0}^{\infty} S(\omega) d\omega = (\text{rms value})^2$$

These relations, known as "Wiener-Khinchine relations".

4. **Quadspectrum-Cospectrum**

The cross-PSD $S_{XY}(\omega)$ can be written as

$$S_{XY}(\omega) = \int_{-\infty}^{+\infty} R_{XY}(\tau) e^{-i\omega\tau} d\tau = C_{XY}(\omega) - iQ_{XY}(\omega) \tag{4.35}$$

where,

$$C_{XY}(\omega) = \int_{-\infty}^{+\infty} R_{XY}(\tau)\cos(\omega\tau) d\tau$$

is called the cospectrum or coincident spectral density, and

$$Q_{XY}(f) = \int_{-\infty}^{+\infty} R_{XY}(\tau)\sin(\omega\tau)\mathrm{d}\tau$$

is called the quadspectrum or quadrature spectral density function.

4.4 Several Typical Random Processes and Their Characteristics

4.4.1 Gaussian Random Process

In many practical scenarios, it is concluded that vibrations are stationary and Gaussian, which simplifies the problem of calculating the response of a system.

Let $X(t)$ be a random process, and consider its 1st and 2nd order probability density functions (PDF).

$$p(x;t) = \frac{1}{\sqrt{2\pi}\sigma_X(t)}\exp\left\{-\frac{1}{2}\left[\frac{x - m_X(t)}{\sigma_X(t)}\right]^2\right\}, \quad -\infty < x < \infty \tag{4.36}$$

$$p(x_1, x_2; t_1, t_2) = \frac{1}{2\pi\sigma_1\sigma_2\sqrt{1 - r_{12}^2}}\times$$

$$\exp\left\{-\frac{1}{2\left(1 - r_{12}^2\right)}\left[\frac{(x_1 - m_1)^2}{\sigma_1^2} + \frac{(x_2 - m_2)^2}{\sigma_2^2} - 2r_{12}\frac{(x_1 - m_1)(x_2 - m_2)}{\sigma_1\sigma_2}\right]\right\}; \tag{4.37}$$

$$-\infty < x_1, x_2 < \infty$$

$$m_1 = m_X(t_1), \quad m_2 = m_X(t_2); \quad \sigma_1 = \sigma_X(t_1), \quad \sigma_2 = \sigma_X(t_2), \quad r_{12} = r_{XX}(t_1,\ t_2)$$

Further, consider time instants $\{t_i\}_{i=1}^n$ and associated random variables $\{X(t_i)\}_{i=1}^n$. Let JPDF of $\{X(t_i)\}_{i=1}^n$ be given by

$$p(x_1, x_2, \ldots, x_n; t_1, t_2, \ldots, t_n) = \frac{1}{(2\pi)^{\frac{n}{2}}\left|S_{ij}\right|^{\frac{1}{2}}}\exp\left[-\frac{1}{2}(x - \eta)^t S_{ij}^{-1}(x - \eta)\right] \tag{4.38}$$

$$-\infty < x_1 < \infty \forall \ i \in [1, n]$$

where

$$S_{ij} = \langle [X(t_i) - m_X(t_i)][X(t_j) - m_X(t_j)] \rangle \tag{4.39}$$

Note: $\boldsymbol{S}^t$ is the covariance matrix $\boldsymbol{S}$ and $\boldsymbol{S}$ is positive definite.

$$\eta = [m_X(t_1) \quad m_X(t_2) \quad \cdots \quad m_X(t_n)]^{\mathrm{T}}$$

$$x = (x_1 \quad x_2 \quad \cdots \quad x_n)^{\mathrm{T}}$$

$X(t)$ is said to be a Gaussian random process if the above PDF holds true for any n and for any choice of $\{t_i\}_{i=1}^n$.

A Gaussian random process has the following characteristics:

(1) A Gaussian random process is completely specified through its mean $m_X(t)$ and covariance $C_{XX}(t_1, t_2)$.
(2) $X(t)$ is stationar y $\Rightarrow$ $m_X(t) = m_X, C_{XX}(t_1, t_2) = C_{XX}(t_2 - t_1) \Rightarrow p_{XX}(x_1, x_2; t_1, t_2) = p_{XX}(x_1, x_2; t_2 - t_1) \Rightarrow X(t)$ is second order strictly stationary stochastic process.
(3) A stationary Gaussian random process with zero mean is completely described by its autocovariance function.
(4) Linear transformation of Gaussian random processes retains the Gaussian nature. Gaussian distributed loads on linear systems yield Gaussian distributed responses.

4.4.2 Poisson Random Process

The Poisson random process is a fundamental independent incremental process that accumulates the number of occurrences of random events. For example, the cumulative number of calls received by a certain telephone exchange over time constitutes a Poisson process.

1. **Definition**

A Poisson process is a type of random process defined by the timing of event occurrences. A random process is a time homogeneous one-dimensional Poisson process if it satisfies the following conditions:

(1) The number of events occurring within two mutually exclusive (non-overlapping intervals is an independent random variable.
(2) The probability distribution of the number of events occurring within the interval $[t, t+\tau]$ is

$$P[N(t+\tau) - N(t) = k] = \frac{\mathrm{e}^{-\lambda\tau}(\lambda\tau)^k}{k!}, \quad k = 0, 1, 2, \ldots \tag{4.40}$$

where, λ is a positive and fixed parameter, commonly referred to as the arrival rate or intensity. Therefore, given the number of events occurring within a time interval, the random variable $[N(t+\tau) - N(t)]$ exhibits a Poisson distribution in $[t, t+\tau]$, with a parameter $\lambda\tau$.

More generally, a Poisson process is a process that assigns a random number of events to each bounded region in each bounded time interval or in a certain space (such as an Euclidean plane or a three-dimensional Euclidean space), such that:

① The number of events in one time interval or spatial region and the number of events in another mutually exclusive (non-overlapping) time interval or spatial region are independent of each other.
② The number of events in each time interval or spatial region is a random variable that follows a Poisson distribution. Technically, more precisely, every set with finite measures is assigned a Poisson distribution random variable.

The Poisson process is one of the most famous processes in the Lévy process. The homogeneous time Poisson process is also an example of a continuous time Markov process with homogeneous time. A homogeneous, one-dimensional Poisson process is a pure birth process and the simplest example of a birth–death process.

2. **Characteristic**

Consider a Poisson process, where we record the arrival time of the first event as T_1. In addition, for $n > 1$, T_n is used to record the time spent between the first and n-th events. The sequence $\{T_n, n = 1, 2, \ldots\}$ is called the arrival interval time series.

$T_n(n = 1, 2, \ldots)$ is an independent and identically distributed exponential random variable with a mean of $1/\lambda$.

4.4.3 Markov Random Process

Markov process is a class of random process. Its original model, the Markov chain, was developed by Russian mathematician A. A. Markov in 1907. Markov process is an essential method for studying the state space of discrete event dynamic systems, and the mathematical foundation is stochastic process theory. People often encounter random processes with the following characteristics in practice: under the condition that the state is known, its future evolution does not depend on its past evolution. The property of "future" and "past" being independent under the known "present" condition is called Markov property, and a stochastic process with this property is called the Markov process.

1. **Markov Sequence**

Markov sequence refers to a Markov process with discrete parameters and continuous states. A random variable sequence $X_n(n = 1, 2, \ldots)$, if for any n, there is

$$F_X(x_n|x_{n-1}, \ldots, x_2, x_1) = F_X(x_n|x_{n-1}) \tag{4.41}$$

or

$$f_X(x_n|x_{n-1},\ldots,x_2,x_1)=f_X(x_n|x_{n-1}) \tag{4.42}$$

X_n is called Markov sequence. The joint probability density of X_n is

$$f_X(x_1,x_2,\ldots,x_n)=f_X(x_n|x_{n-1})f_X(x_{n-1}|x_{n-2})\cdots f_X(x_2|x_1)f_X(x_1) \tag{4.43}$$

Markov sequence has the following features:

(1) The subsequence of a Markov sequence is still a Markov sequence.
(2) The joint probability density of any collection of random variables is the product of their individual probability densities. $f_X(x_n|x_{n+1},x_{n+2},\ldots,x_{n+k})=f_X(x_n|x_{n+1})$.
(3) $E(x_n|x_{n-1},\ldots,x_2,x_1)=E(x_n|x_{n-1})$.
(4) In a Markov sequence, if the present is known, then the future and the past are independent of each other. There is a relationship:

$$f_X(x_n,x_s|x_r)=f_X(x_n|x_r)f_X(x_s|x_r),\quad n>r>s \tag{4.44}$$

(5) If the conditional probability density $f_X(x_n|x_{n-1})$ is independent of n, then the Markov sequence is said to be homogeneous.
(6) If a Markov sequence is homogeneous and all random variables X_n have the same probability density, then the Markov sequence is said to be stationary.
(7) The transition probability density of a Markov sequence satisfies the Chapman-Kolmogorov equation, which is

$$f_X(x_n|x_s)=\int_{-\infty}^{\infty} f_X(x_n|x_r)f_X(x_r|x_s)\mathrm{d}x_r,\quad n>r>s \tag{4.45}$$

2. **Markov Chain**

(1) The Markov Chain Definition

Assume that $\boldsymbol{X}_n(n=1,2,\ldots)$ is a discrete-time stochastic process with a state space $\boldsymbol{I}=\{a_1,a_2,\ldots,a_N\}$. If the probability of the process being in any state $a_{i_{m+k}}(i=1,2,\ldots,N)$ at time t_{m+k} is only related to the state of the process at t_m, and is independent of the state of the process before t_m,

$$P\{X_{m+k}=a_{i_{m+k}}|X_m=a_{i_m},\ldots,X_1=a_{i_1}\}=P\{X_{m+k}=a_{i_{m+k}}|X_m=a_{i_m}\} \tag{4.46}$$

then the process is called a Markov chain.

(2) The Transition Probability and Finite Dimensional Distribution of Markov Chains

Its definition is:

$$p_{ij}(m, m+k) = P\{X_{m+k} = a_j | X_m = a_i\} \tag{4.47}$$

where, $i, j = 1, 2, \ldots, N$; m, k are positive integers.

If $p_{ij}(m, m+k)$ is independent of m, it is called a homogeneous Markov chain. The Markov chains discussed later are all homogeneous Markov chains.

The transition probability of a Markov chain and its matrix are defined as

$$p_{ij} = p_{ij}(1) = p_{ij}(m, m+1) = P\{X_{m+1} = a_j | X_m = a_i\} \tag{4.48}$$

$$\boldsymbol{P} = \boldsymbol{P}(1) = \begin{pmatrix} p_{11} & p_{12} & \cdots & p_{1N} \\ p_{21} & p_{22} & \cdots & p_{2N} \\ \vdots & \vdots & & \vdots \\ p_{N1} & p_{N2} & \cdots & p_{NN} \end{pmatrix} \tag{4.49}$$

The one-step transition probability matrix $\boldsymbol{P}$ has the following two properties:

$$0 \le p_{ij} \le 1 \tag{4.50}$$

$$\sum_{j=1}^{N} p_{ij} = 1 \tag{4.51}$$

The high-order transition probability and its matrix of Markov chains are defined as

$$p_{ij}(n) = p_{ij}(m, m+n) = P\{X_{m+n} = a_j | X_m = a_i\} \tag{4.52}$$

$$\boldsymbol{P}(n) = \begin{pmatrix} p_{11}(n) & p_{12}(n) & \cdots & p_{1N}(n) \\ p_{21}(n) & p_{22}(n) & \cdots & p_{2N}(n) \\ \vdots & \vdots & & \vdots \\ p_{N1}(n) & p_{N2}(n) & \cdots & p_{NN}(n) \end{pmatrix} \tag{4.53}$$

The n-step transition probability matrix $\boldsymbol{P}(n)$ has the following two properties:

$$0 \le p_{ij}(n) \le 1 \tag{4.54}$$

$$\sum_{j=1}^{N} p_{ij}(n) = 1 \tag{4.55}$$

In addition, it is also stipulated that

$$p_{ij}(0) = p_{ij}(m, m) = \delta_{ij} = \begin{cases} 1 & i = j \\ 0 & i \neq j \end{cases}$$

The n-step transition probability of a Markov chain and its matrix have the following discrete form of the Chapman-Kolmogorov equation:

$$p_{ij}(n) = p_{ij}(l + k) = \sum_{r=1}^{N} p_{ir}(l) p_{rj}(k) \tag{4.56}$$

$$P(n) = P(l + k) = P(l)P(k) \tag{4.57}$$

When n is any positive integer, there is

$$P(n) = PP(n - 1) = \cdots = P^n \tag{4.58}$$

In Eq. (4.57), if $n = k + 1$, there has

$$p_{ij}(k + 1) = \sum_{r=1}^{N} p_{ir} p_{rj}(k) = \sum_{r=1}^{N} p_{ir}(k) p_{rj} \tag{4.59}$$

From the above, we know that the probability rule of the Markov chain state transition process is determined by the one-step transition probability matrix $\boldsymbol{P}$ with the one-step transition probability p_{ij} as the element. However, $\boldsymbol{P}$ cannot determine the initial probability distribution, so it is necessary to introduce the initial probability:

$$p_i = P\{X_0 = a_i\}, \quad i = 0, 1, 2, \ldots \tag{4.60}$$

And $\{p_i\} = p_0, p_1, p_2, \ldots,$ is referred to as the initial distribution, obviously has

$$0 \leq p_i \leq 1, \quad \sum_i p_i = 1 \tag{4.61}$$

If the absolute probability $p_j(k) = P\{X_k = a_j\}$, then there is

$$p_j(k + 1) = \sum_i p_i p_{ij}(k + 1) = \sum_i p_i(k) p_{ij} \tag{4.62}$$

The finite dimensional distribution of Markov chains can be expressed as

$$P\{X_0 = a_{i_0}, X_1 = a_{i_1}, \ldots, X_n = a_{i_n}\} = P\{X_0 = a_{i_0}\}P\{X_1 = a_{i_1}|X_0 = a_{i_0}\}$$

$$\cdots P\{X_n = a_{i_n}|X_{n-1} = a_{i_{n-1}}\}$$

$$= p_{i_0}p_{i_0 i_1}\cdots p_{i_{n-1}i_n} \tag{4.63}$$

3. **Classification of States in Markov Chains**

The purpose of introducing the concept of state period in the inference is to study the slimit properties of the state transition matrix, that is, the limit of n $\boldsymbol{P}(n)$ when n approaches infinity, which can reflect the characteristics of the Markov chain in a stationary state.

(1) Arrival and Communication

Reachability: Let i and j be two states of a time-homogeneous Markov chain. If there exists an integer $n \geq 0$ exists, such that $p_{ij}^{(n)} > 0$, then it is said that state j can be reached from state i, denoted as $i \to j$. Conversely, if $p_{ij}^{(n)} = 0$ for all $n \geq 0$, state j represents an unreachable state from state i.

If $i \to j$ and $j \to i$ hold, then states i and j are considered to reach each other (connected), denoted as $i \leftrightarrow j$.

Mutual accessibility is an equivalence relation that satisfies the following properties:

① Reflexivity: $i \leftrightarrow i$;
② Symmetry: If $i \leftrightarrow j$, then $j \leftrightarrow i$;
③ Transitivity: If $i \leftrightarrow k$ and $k \leftrightarrow j$, then $i \leftrightarrow j$.

Equivalence relations can be used to partition sets, therefore, reachability can also classify state spaces into equivalent classes under the reachability relation.

Irreducible state space: the state space of a Markov chain is deemed irreducible if it belongs to the same class under the reachability equivalence relation. If this is not the case, the Markov chain is referred to as reducible.

Period of a state: let i be a state of the Markov chain. The greatest common divisor of all positive integers $n(n \geq 1)$ for which $p_{ii}^{(n)} > 0$ is called the period of state i, denoted as $\mathrm{d}(i)$ or $\mathrm{d}i$.

If $p_{ii}^{(n)} = 0$ for all $n \geq 1$, the period is agreed to be ∞; and the state i with $\mathrm{d}(i) = 1$ is termed non-periodic. It should be noted that when the period of state state i with $\mathrm{d}(i) = 1$ may not hold true.

Moreover, if n cannot be evenly divided by the period $\mathrm{d}(i)$, then $p_{ii}^{(n)} = 0$ must hold true.

(2) Frequent and Non-recurrent States (or Transient States)

For any two states i and j, introduce a random variable $\{X_0 = i\}$ on the event

$$T_{ij} = \min\{n : X_0 = i, X_n = j; n \geq 1\} \tag{4.64}$$

where, T_{ij} is the moment when the system first enters state j starting from state i. Alternatively, or more precisely, $\omega \in \{X_0 \neq i\}$ or $\omega \in \{X_0 = i, X_n \neq j; n \geq 1\}$, can be supplemented as follows:

$$T_{ij}(\omega) = \infty \tag{4.65}$$

The definition can be further elaborated as

$$f_{ij}(n) = P\{T_{ij} = n | X_0 = i\} \geq 0 \tag{4.66}$$

It is the probability of the system reaching state j for the first time after n steps of transition from state i. Obviously, there is

$$\begin{aligned} f_{ij} &= P\{X_n = j, X_k \neq j, k = 1, \ldots, n-1 | X_0 = i\} \\ &= \sum_{i_1 \neq j} \cdots \sum_{i_{n-1} \neq j} p_{ii_1} p_{i_1 i_2} \cdots p_{i_{n-1} j} (n \geq 1) \end{aligned} \tag{4.67}$$

hence

$$f_{ij}(1) = p_{ij} = P\{X_1 = j | X_0 = i\}$$

$$f_{ij}(\infty) = P\{X_k \neq j, k \geq 1 | X_0 = i\}$$

give

$$f_{ij} = \sum_{1 \leq n \leq \infty} f_{ij}(n) = \sum_{1 \leq n \leq \infty} P\{T_{ij} = n | X_0 = i\} = P\{T_{ij} < \infty\} \tag{4.68}$$

Consequently, f_{ij} represents the probability of starting from state i and ultimately reaching state j

Property: when $i \neq j$, then $i \to j$, $\quad f_{ij} > 0$. If $f_{ij} = 1$, state i is said to be recurrent. Otherwise, if $f_{ij} < 1$, state i is referred to as non-recurrent or transient. If state i is recurrent, the probability of starting from state i and transitioning through a finite number of steps before finally returning to $i = 1$.

(3) Ergodicity and Stationary Distribution

① Ergodicity

Let $X(n)$ be a homogeneous Markov chain. If for all states i and j, there exists a limit independent of i

$$\lim_{n \to \infty} p_{ij}(n) = p_i \tag{4.69}$$

the Markov chain $X(n)$ is said to exhibit ergodicity.

For a homogeneous Markov chain $X(n)$ with a finite number of states, if there exists a positive integer m such that

$$p_{ij}(m) > 0, \quad i, j = 1, 2, \ldots, N$$

then the chain is ergodic. In Eq. (4.41), $\{p_i\} = \{p_1, p_2, \ldots, p_N\}$ is the unique solution,

$$p_i = \sum_{i=1}^{N} p_i p_{ij}, \quad j = 1, 2, \ldots, N \tag{4.70}$$

when the conditions are met

$$0 < p_i < 1, \quad \sum_{j=1}^{N} p_j = 1$$

② Stationary Distribution

A probability distribution $\{v_i\}$ of Markov chain,$v_i \geq 0$ and $\sum_{j=0}^{N} v_j = 1$, if

$$v_j = \sum_{i=0}^{N} v_i p_{ij}$$

It is called the stationary distribution of the chain. And there is

$$v_j = \sum_{i=0}^{N} v_i p_{ij}(n)$$

where, N is the total number of states.

4. **Markov Process**

The Markov process discussed here refers to a process that is continuous in both time and state space. The diffusion process is a special case of this type of Markov process.

Let $\{X(t), t \in T\}$ be a random process, where E is its state space. If the conditional distribution function of any $t_1 < t_2 < \cdots < t_n < t$, and $x_1, x_2, \ldots, x_n \in E$, with $X(t) \in T$, is only related to $X(t_n) = x_n$ and independent of $X(t_1) = x_1, \ldots, X(t_{n-1}) = x_{n-1}$, that is, the conditional distribution function satisfies the equation

$$F_X(x_n; t_n | x_{n-1}, \ldots, x_2, x_1; t_{n-1}, \ldots, t_2, t_1) = F_X(x_n; t_n | x_{n-1}; t_{n-1})(n \geq 3) \tag{4.71}$$

namely,

$$P\{X(t_n) \le x_n | X(t_{n-1}) \le x_{n-1}, X(t_{n-2}) \le x_{n-2}, \ldots, X(t_1) \le x_1\}$$

$$= P\{X(t_n) \le x_n | X(t_{n-1}) \le x_{n-1}\} \tag{4.72}$$

then $X(t)$ is said to possess the Markov property, also known as the no-aftereffect or no-memory property.

If $X(t)$ is a discrete random variable, then the Markov property also satisfies

$$P\{X(t) = x | X(t_n) = x_n, \ldots, X(t_1) = x_1\} = P\{X(t) = x | X(t_n) = x_n\} \tag{4.73}$$

For any n and any choice of $t_1 < t_2 < \cdots < t_n$, we can deduce

$$P_X(x_n, t_n | x_{n-1}, t_{n-1}; x_{n-2}, t_{n-2}; \cdots; x_1, t_1)$$

$$= P_X(x_n, t_n | x_{n-1}, t_{n-1}) p_X(x_n, t_n | x_{n-1}, t_{n-1}; x_{n-2}, t_{n-2}; \ldots; x_1, t_1)$$

$$= p_X(x_n, t_n | x_{n-1}, t_{n-1}) \tag{4.74}$$

The transition probability distribution of the Markov process is defined as

$$F_X(x_n; t_n | x_{n-1}; t_{n-1}) = P\{X(t_n) \le x_n | X(t_{n-1}) \le x_{n-1}\} \tag{4.75}$$

or

$$F_X(x; t | x_0; t_0) = P\{X(t) \le x | X(t_0) \le x_0\}, \quad t > t_0 \tag{4.76}$$

The transition probability distribution is a distribution function about x, therefore, there exists

(1) $F_X(x; t|x_0; t_0) \ge 0$.
(2) $F_X(\infty; t|x_0; t_0) = 1$.
(3) $F_X(-\infty; t|x_0; t_0) = 0$.
(4) $F_X(x; t|x_0; t_0)$ is about x monotonically non decreasing, right continuous functions.
(5) Satisfy Chapman-Kolmogorov equation.

$$F_X(x; t | x_0; t_0) = \int_{-\infty}^{\infty} F_X(x; t | x_1; t_1) F_X(x_1; t_1 | x_0; t_0) \mathrm{d}x_1 \tag{4.77}$$

The transition probability density of Markov process is defined as

$$f_X(x; t | x_0; t_0) = \frac{\partial}{\partial x} F_X(x; t | x_0; t_0) \tag{4.78}$$

hence, there exists

$$\int_{-\infty}^{\infty} f_X(x; t|x_1; t_1)\mathrm{d}x = 1 \tag{4.79}$$

$$f_X(x; t|x_0; t_0) \to \delta(x - x_0), \quad t \to t_0 \tag{4.80}$$

$$f_X(x_n; t_n \mid x_{n-1}, \ldots, x_2, x_1; t_{n-1}, \ldots, t_2, t_1) = f_X(x_n; t_n \mid x_{n-1}; t_{n-1}), \quad n \geq 3 \tag{4.81}$$

The transition probability density of the Markov process satisfies the Chapman-Kolmogorov equation.

$$f_X(x_n; t_n|x_k; t_k) = \int_{-\infty}^{\infty} f_X(x; t|x_r; t_r) f_X(x_r; t_r|x_k; t_k)\mathrm{d}x, \quad t_k < t_r < t_n \tag{4.82}$$

If the Markov process $X(t)$ has

$$F_X(x; t|x_0; t_0) = F_X(x|x_0; \tau), \quad \tau = t - t_0 \tag{4.83}$$

or

$$f_X(x; t|x_0; t_0) = f_X(x|x_0; \tau), \quad \tau = t - t_0 \tag{4.84}$$

it is called a homogeneous Markov process.

The n-dimensional probability density of Markov process $X(t)$ can be written as

$$f_X(x_1, x_2, \ldots, x_{n-1}, x_n; t_1, t_2, \ldots, t_{n-1}, t_n)$$

$$= f_X(x_1; t_1) \prod_{i=1}^{n-1} f_X(x_{i+1}; t_{i+1}|x_i; t_i), \quad t_1 < t_2 < \cdots < t_n \tag{4.85}$$

The following processes are common examples of Markov Process:

① Independent random process is Markov process.
② An independent increment process is a Markov process: let $\{X(t), t \in [0, +\infty)\}$ be an independent increment process, and if $P(X(0) = x_0) = 1,\ x_0$ is a constant, then $X(t)$ is a Markov process.
③ The Poisson process is a Markov process.
④ The Wiener process is a Markov process.
⑤ The random walk process of particles is a Markov process.

4.5 Differentiation and Integration of Random Processes

In the formulation of problems in mechanics, it is often necessary to differentiate random processes. For instance, let $X(t)$ be a random process representing displacement, we would be interested in its velocity and acceleration.

For deterministic functions,

$$\frac{\mathrm{d}z}{\mathrm{d}t} = \lim_{\Delta \to 0} \frac{z(t+\Delta) - z(t)}{\Delta} \tag{4.86}$$

By selecting a sequence of Δ_s of the form $\{\Delta_i\}_{i=1}^{n}$, such that $\Delta_{i+1} < \Delta_i$, we obtain a sequence of numbers $y_i = \frac{z(t+\Delta_i)-z(t)}{\Delta_i}$ and we seek to determine $\lim\limits_{i\to\infty} y_i$.

When $X(t)$ is a random process, the sequence

$$Y_i = \frac{X(t+\Delta_i) - X(t)}{\Delta_i}, \quad i = 1, 2, \ldots \tag{4.87}$$

where, Y_i is a sequence of random variable.

There are several valid modes of convergence for random variables. Consequently, the associated calculus would also be constructed based on a chosen mode of convergence for random variables.

Definition: a sequence of random variables $X_1, X_2, \ldots, X_n$, is said to converge to the random variable X in the mean square sense if

$$\lim_{n\to\infty} \left\langle (X - X_i)^2 \right\rangle \to 0 \tag{4.88}$$

This is denoted by $\lim_{n\to\infty} X_i \to X$. The calculus based on this definition of convergence of random variables is called the mean square calculus. This leads to the definition of the mean square derivative and mean integral.

Consider:

$$\begin{aligned} \left\langle X(t_1)X'(t_2)\right\rangle &= \left\langle X(t_1) \lim_{\Delta\to\infty} \frac{X(t_2+\Delta) - X(t_2)}{\Delta} \right\rangle \\ &= \lim_{\Delta\to\infty} \frac{\langle X(t_1)X(t_2+\Delta)\rangle - \langle X(t_1)X(t_2)\rangle}{\Delta} \\ &= \lim_{\Delta\to\infty} \frac{R_{XX}(t_1, t_2+\Delta) - R_{XX}(t_1, t_2)}{\Delta} = \frac{\partial R_{XX}(t_1, t_2)}{\partial t_2} \end{aligned} \tag{4.89}$$

The autocorrelation function is

$$R_{XX}(t_1, t_2) = \frac{\partial R_{XX}(t_1, t_2)}{\partial t_2} \tag{4.90}$$

Similarly, it can be shown that

$$\langle X'(t_1)X'(t_2)\rangle = R_{XX}(t_1, t_2) = \frac{\partial^2 R_{XX}(t_1, t_2)}{\partial t_1 \partial t_2} \tag{4.91}$$

And, more generally

$$\left\langle \left.\frac{\mathrm{d}^n X(t)}{\mathrm{d}t^n}\right|_{t=t_1} \left.\frac{\mathrm{d}^m X(t)}{\mathrm{d}t^m}\right|_{t=t_2} \right\rangle = \frac{\partial^{n+m} R_{XX}(t_1, t_2)}{\partial t_1^n \partial t_2^m} \tag{4.92}$$

Exercise 4

E4.1: The sample functions of a certain state ergodic a random process are all square waves with an amplitude equal to a and a period of T, as shown in Fig. 4.10. The initial phase of each sample function is a uniformly distributed random variable between 0 and T. Known a, T, find $R_x(\tau)$.

E4.2: Find the autocorrelation function of the narrow band random process which is given in Fig. 4.11. Known S_0, ω_1, ω_2, and $|\omega_1 - \omega_2| \ll \omega_1$, find $R_x(\tau)$.

E4.3: Supposing a sample function of an ergodic random process is rectangle function, see in Fig. 4.12.

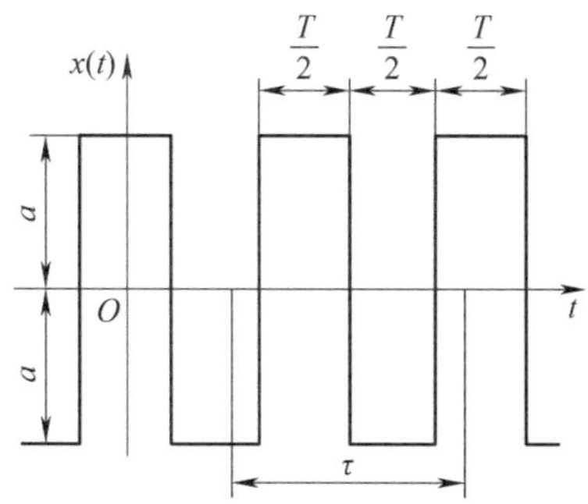

Fig. 4.10 A square wave signal

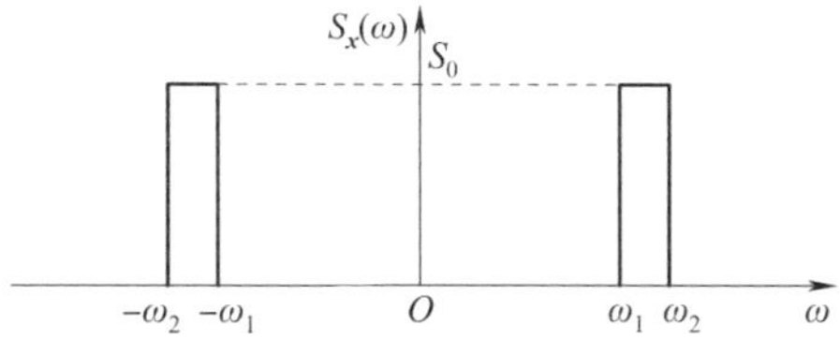

Fig. 4.11 A sketch of a potential energy barrier

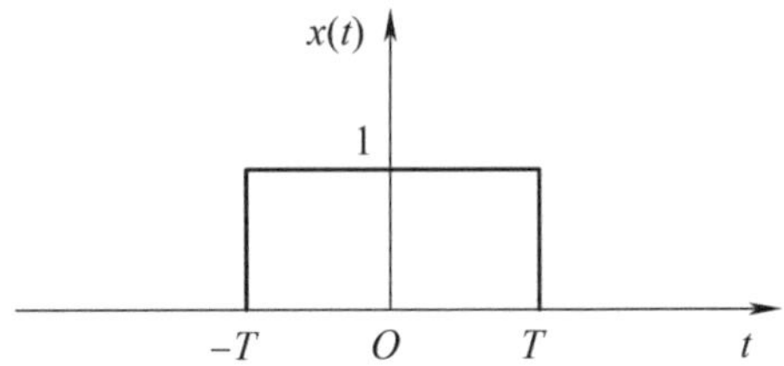

Fig. 4.12 A rectangular pulse waveform

$$x(t) = \begin{cases} 1 & |t| \leq T \\ 0 & |t| > T \end{cases}$$

Find this function's PSD $S_x(\omega)$.

E4.4: Let $U(t)$ and $V(t)$ be defined as $U(t) = S(t)$, $V(t) = S(t) + W(t)$. $S(t)$: stationary Gaussian random process with zero mean.$W(t)$: zero mean Gaussian white noise independent of $S(t)$ with $\langle W(t)W(t+\tau)\rangle = 2\pi\delta(\tau)$. Determine $Cov_{UV}(\omega)$.

Chapter 5
SDOF System

This chapter begins by introducing the mechanical vibration of a SDOF linear system. It outlines the modeling and analysis of engineering problems that can be approximated using a SDOF system. Building upon the established mathematical model for the SDOF system, this chapter delves into the system's response under various random excitations.

5.1 SDOF System Modeling

To analyze the motion of a SDOF system, it is essential to construct a mathematical description that captures its dynamic characteristics. This description is referred to as the model and can be derived from the application of known physical laws.

Consider a SDOF system force diagram as depicted in Fig. 5.1.

The magnitude of the restoring force is given by Eq. (5.1).

$$|F_{\mathrm{r}}| = k|x| \tag{5.1}$$

If $x > 0$, F_{r} is opposite to the positive direction of $x -$ axis.$F_{\mathrm{r}} = -k|x| = -kx$.

If $x > 0$, F_{r} has the same direction as x-axis. $F_{\mathrm{r}} = k|x| = -kx$.

Therefore, F_{r} can be represented in

$$F_{\mathrm{r}} = -kx \tag{5.2}$$

A similar analysis leads to the conclusion that the damping force can be expressed by Eq. (5.3).

$$F_{\mathrm{d}} = -c\dot{x} \tag{5.3}$$

J. Zhang and H. Zhou, *Random Vibration and Spectrum Analysis*,
https://doi.org/10.1007/978-981-95-0191-5_5

Fig. 5.1 A SDOF system force diagram

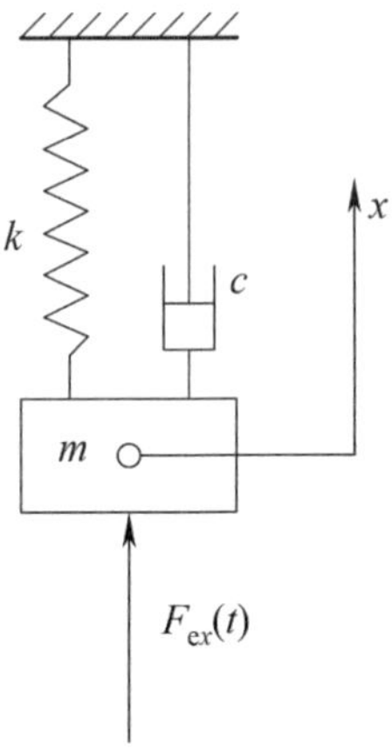

Assuming the system is linear, the motion equation of the SDOF system is given by Eq. (5.4).

$$\begin{aligned} &m\ddot{x} + c\dot{x} + kx = F_{ex}(t) \\ &x(0) = x_0, \quad \dot{x}(0) = \dot{x}_0 \end{aligned} \tag{5.4}$$

The external force $F_{ex}(t)$ will be denoted by $f(t)$ in subsequent discussions.

5.1.1 Forced Vibration Under Harmonic Input

When the input is

$$f(t) = P\cos(\lambda t) \tag{5.5}$$

Derive the kinematic equation as per equation

$$m\ddot{x} + c\dot{x} + kx = P\cos(\lambda t) \tag{5.6}$$

The initial conditions are $x(0) = x_0, \ \dot{x}(0) = \dot{x}_0$, specified by equation

$$\ddot{x} + 2\eta\omega\dot{x} + \omega^2 x = (P/m)\cos(\lambda t) \tag{5.7}$$

where, m is the mass, k is the stiffness, and η is the damping ratio, $\omega^2 = \frac{k}{m}, \quad \eta = \frac{c}{2\sqrt{mk}}$.

Example 5.1 Given a system with $m = 10\,\text{kg},\ k = 1\ 000\,\text{N/m},\ \eta = 0.05,\ x_0 = 0.15$, calculate its response.

Solution:

$$x(t) = x_{\text{cf}}(t) + x_{\text{pi}}(t)$$

$$x_{\text{cf}}(t) = \exp(-\eta\omega t)[A\cos(\omega_{\text{d}}t) + B\sin(\omega_{\text{d}}t)]$$

$$x_{\text{pi}}(t) = C\cos(\lambda t) + D\sin(\lambda t)$$

$$C = \frac{(P/m)(\omega^2 - \lambda^2)}{(\omega^2 - \lambda^2)^2 + (2\eta\omega\lambda)^2},\quad D = \frac{(P/m)2\eta\omega\lambda}{(\omega^2 - \lambda^2)^2 + (2\eta\omega\lambda)^2}$$

$$x(t) = \exp(-\eta\omega t)[A\cos(\omega_{\text{d}}t) + B\sin(\omega_{\text{d}}t)] + \frac{(P/m)\cos(\lambda t - \theta)}{\left[(\omega^2 - \lambda^2)^2 + (2\eta\omega\lambda)^2\right]^{\frac{1}{2}}}$$

$$x_0 = A + \frac{(P/m)\cos\theta}{\left[(\omega^2 - \lambda^2)^2 + (2\eta\omega\lambda)^2\right]^{\frac{1}{2}}}$$

$$\dot{x}_0 = -\eta\omega A + B\omega_{\text{d}} + \frac{(P/m)\sin\theta}{\left[(\omega^2 - \lambda^2)^2 + (2\eta\omega\lambda)^2\right]^{\frac{1}{2}}}$$

Nature of steady state response

$$\begin{aligned}\lim_{t\to\infty} x(t) &= \frac{(P/m)\cos(\lambda t - \theta)}{\left[(\omega^2 - \lambda^2)^2 + (2\eta\omega\lambda)^2\right]^{\frac{1}{2}}}\\ &= \frac{(P/k)\cos(\lambda t - \theta)}{\left[(1 - r^2)^2 + (2\eta r)^2\right]^{\frac{1}{2}}}\\ &= X\cos(\lambda t - \theta)(r = \lambda/\omega,\quad \omega = \sqrt{k/m})\end{aligned}$$

$$\frac{X}{P/k} = \text{DMF} = \frac{1}{\left[(1 - r^2)^2 + (2\eta r)^2\right]^{\frac{1}{2}}}$$

$$\theta = \arctan\left[\frac{2\eta r}{(1 - r^2)}\right]$$

where, DMF is the dynamic magnification factor, *P*/*k* is the static response under force *P*.

5.1.2 Resonance in Undamped System

Consider the following system under periodic excitation as Eq. (5.8).

$$\ddot{x} + \omega^2 x = (P/m)\cos(\lambda t) \tag{5.8}$$

Get

$$x(t) = A\cos(\omega_{\mathrm{d}}t) + B\sin(\omega_{\mathrm{d}}t) + \frac{(P/m)\cos(\lambda t)}{(\omega^2 - \lambda^2)} \tag{5.9}$$

Initial conditions are

$$x(0) = 0, \quad \dot{x}(0) = 0$$

$$x(t) = (P/m)\frac{\cos(\lambda t) - \cos(\omega t)}{\omega^2 - \lambda^2}$$

$$\lim_{\lambda \to \omega} x(t) \to \frac{Pt}{2m\omega}\sin(\omega t)$$

$$\lim_{\substack{\lambda \to \omega \\ t \to \infty}} x(t) \to \infty.$$

5.1.3 Response to an Indicial Function

When the excitation is an indicial function, the response is determined by Eq. (5.10).

$$m\ddot{x} + c\dot{x} + kx = U(t) \tag{5.10}$$

Initial conditions are

$$x(0) = 0, \quad \dot{x}(0) = 0$$

The response is

$$\begin{aligned} x(t) &= \exp(-\eta\omega t)[A\cos(\omega_{\mathrm{d}}t) + B\sin(\omega_{\mathrm{d}}t)] + \frac{1}{k} \\ &= \frac{1}{k}\left\{1 - \exp(-\eta\omega t)\left[\cos(\omega_{\mathrm{d}}t) + \frac{\eta}{\sqrt{1-\eta^2}}\sin(\omega_{\mathrm{d}}t)\right]\right\} \end{aligned} \tag{5.11}$$

5.1.4 Response to a Box Function

When the excitation is a box function, the response is given by Eq. (5.12).

$$m\ddot{x} + c\dot{x} + kx = U(t) - U(t - a) \tag{5.12}$$

Initial conditions are

$$x(0) = 0, \quad \dot{x}(0) = 0$$

$$x(t) = G(t), \quad t < a$$

$$x(t) = G(t) - G(t - a), \quad t > a.$$

5.1.5 Response to a Unit Impulse

1. **Dirac's Delta Function**

$$\delta(t - a) = 0, \quad t \neq a$$

$$\int_{-\infty}^{\infty} f(t)\delta(t - a)\mathrm{d}t = f(a)$$

$$\int_{-\infty}^{\infty} \delta(t - a)\mathrm{d}t = 1 \tag{5.13}$$

2. **Response to a Unit Impulse at $t = 0$**

When the excitation is a unit impulse, the response is

$$f(t) = F[U(t) - U(t - T_\mathrm{c})]$$

$$x(t) = FG(t), \quad t < T_\mathrm{c}$$

$$x(t) = F[G(t) - G(t - T_\mathrm{c})], \quad t > T_\mathrm{c}$$

$$f(t) = FT_{\mathrm{c}} \frac{[U(t) - U(t - T_{\mathrm{c}})]}{T_{\mathrm{c}}}$$

$$\lim_{\substack{T_{\mathrm{c}} \to 0 \\ FT_{\mathrm{c}} \to 1}} f(t) = \frac{\mathrm{d}U}{\mathrm{d}t} = \delta(t) \tag{5.14}$$

$$x(t) = FT_{\mathrm{c}} \frac{[G(t) - G(t - T_{\mathrm{c}})]}{T_{\mathrm{c}}}$$

$$\lim_{\substack{T_{\mathrm{c}} \to 0 \\ FT_{\mathrm{c}} \to 1}} x(t) = \frac{\mathrm{d}G}{\mathrm{d}t} = h(t) \tag{5.15}$$

3. **Impulse Response Function**

From Eq. (5.11), we get

$$h(t) = \frac{\mathrm{d}G}{\mathrm{d}t} = \frac{1}{m\omega_{\mathrm{d}}} e^{-\eta\omega t} \sin(\omega_{\mathrm{d}} t) \tag{5.16}$$

Note: the effect of applying an impulse at $t = 0$ is equivalent to imparting an initial velocity at $t = 0$.

$$m\ddot{h} + c\dot{h} + kh = 0$$

$$h(0) = 0, \quad \dot{h}(0) = 1/m$$

$$h(t) = \mathrm{e}^{-\eta\omega t}[A\cos(\omega_{\mathrm{d}} t) + B\sin(\omega_{\mathrm{d}} t)]$$

$$h(0) = 0 \Rightarrow A = 0$$

$$h(t) = B\mathrm{e}^{-\eta\omega t} \sin(\omega_{\mathrm{d}} t)$$

$$\dot{h}(t) = B(-\eta\omega)\mathrm{e}^{-\eta\omega t} \sin(\omega_{\mathrm{d}} t) + B\mathrm{e}^{-\eta\omega t} \omega_{\mathrm{d}} \cos(\omega_{\mathrm{d}} t)$$

$$\dot{h}(0) = \frac{1}{m} = B\omega_{\mathrm{d}}$$

$$\Rightarrow h(t) = \frac{1}{m\omega_{\mathrm{d}}} \mathrm{e}^{-\eta\omega t} \sin(\omega_{\mathrm{d}} t)$$

Generalization: n-th order differential equation is

$$\frac{\mathrm{d}^n h}{\mathrm{d}^n} + \alpha_{n-1}\frac{\mathrm{d}^{n-1} h}{\mathrm{d}t^{n-1}} + \cdots + \alpha_1 \frac{\mathrm{d}h}{\mathrm{d}t} + \alpha_0 h = 0 \tag{5.17}$$

with the initial conditions

$$h(0) = 0, \quad \frac{\mathrm{d}h}{\mathrm{d}t}(0) = 0, \ldots, \quad \frac{\mathrm{d}^{n-2} h}{\mathrm{d}t^{n-2}}(0) = 0, \quad \frac{\mathrm{d}^{n-1} h}{\mathrm{d}t^{n-1}}(0) = 1.$$

5.2 Duhamel's Integral and Response to Arbitrary Excitation

5.2.1 Duhamel's Integral

The kinematic equation of a linear SDOF system under time-varying external load $f(t)$ and viscous damping is a second order ordinary differential equation (ODE), expressed by Eq. (5.18).

$$m\ddot{x} + c\dot{x} + kx = f(t) \tag{5.18}$$

where, m, k, and c represent the system's mass, stiffness, and damping factor, respectively. If the system, initially at rest in equilibrium, is subjected to a unit impact load at time $t = 0$, e.g., $f(t)$ is a Dirac δ-function $\delta(t)$. The system response (referred to as the unit impulse response function) can be solved as Eq. (5.19).

$$h(t) = \begin{cases} \frac{1}{m\omega_\mathrm{d}} \mathrm{e}^{-\eta\omega_n t} \sin(\omega_\mathrm{d} t) & t > 0 \\ 0 & t \le 0 \end{cases} \tag{5.19}$$

where, $\eta = c/(2m\omega_n)$ is the damping ratio of the system, and ω_n is the natural circular frequency. At any moment t, the pulse response under impact load is given by Eq. (5.20).

$$h(t-\tau) = \frac{1}{m\omega_\mathrm{d}} \mathrm{e}^{-\eta\omega_n (t-\tau)} \sin[\omega_\mathrm{d}(t-\tau)], \quad t \ge \tau \tag{5.20}$$

From Eq. (5.18), assuming its initial conditions are $x(0) = x_0, \quad \dot{x}(0) = \dot{x}_0$, we have

$$x(t) = x_\mathrm{cf}(t) + x_\mathrm{pi}(t)$$

$$\mathrm{d}x(t) = h(t-\tau) f(\tau) \mathrm{d}\tau$$

$$x_{\mathrm{pi}}(t) = \int_0^t h(t-\tau)f(\tau)\mathrm{d}\tau$$

$$x(t) = \mathrm{e}^{-\eta\omega t}[A\cos(\omega_{\mathrm{d}}t) + B\sin(\omega_{\mathrm{d}}t)] + \int_0^t h(t-\tau)f(\tau)\mathrm{d}\tau$$

$$x(0) = x_0 = A$$

$$\dot{x}(t) = -\eta\omega\mathrm{e}^{-\eta\omega t}[A\cos(\omega_{\mathrm{d}}t) + B\sin(\omega_{\mathrm{d}}t)] + \mathrm{e}^{-\eta\omega t}[-A\omega_{\mathrm{d}}\sin(\omega_{\mathrm{d}}t) + B\omega_{\mathrm{d}}\cos(\omega_{\mathrm{d}}t)] + \frac{\mathrm{d}}{\mathrm{d}t}\int_0^t h(t-\tau)f(\tau)\mathrm{d}\tau$$

Digress

$$\frac{\mathrm{d}}{\mathrm{d}x}\int_{g(x)}^{q(x)} f(x,\tau)\mathrm{d}\tau = \int_{g(x)}^{q(x)} \frac{\partial f(x,\tau)}{\partial x}\mathrm{d}\tau + \frac{\mathrm{d}q}{\mathrm{d}x}f(x,q(x)) - \frac{\mathrm{d}g}{\mathrm{d}x}f(x,g(x))$$

From initial conditions, get

$$B = \frac{\dot{x}_0 + \eta\omega x_0}{\omega\sqrt{1-\eta^2}} \tag{5.21}$$

$$x(t) = \mathrm{e}^{-\eta\omega t}\left[x_0\cos(\omega_d t) + \frac{\dot{x}_0 + \eta\omega x_0}{\omega\sqrt{1-\eta^2}}\sin(\omega_d t)\right] + \int_0^t h(t-\tau)f(\tau)\mathrm{d}\tau \tag{5.22}$$

Duhamel's integral and response to arbitrary excitation:approximate $f(t)$ as a train of impulses; $\mathrm{d}x(t)$ is response at t due to a single impulse at $t = \tau$ of magnitude $f(\tau)$ dr; $x(t)$ is total response due to all the impulses. For systems starting from rest, Duhamel's integral provides a complete solution.

Example 5.2 A SDOF system is excited by the force as shown in Fig. 5.2. Assume that the system starts from rest. Write down the expression for the response valid for any time *t*.

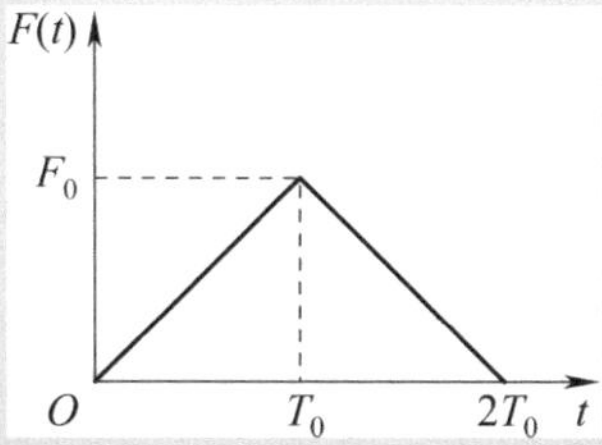

Fig. 5.2 Triangular excitation force

Solution:

From Fig. 5.2, it is known that

$$F(t) = \begin{cases} \frac{F_0}{T_0}t & 0 < t \le T_0 \\ -\frac{F_0}{T_0}t + 2F_0 & T_0 < t \le 2T_0 \\ 0 & t > 2T_0 \end{cases}$$

Get

$$x(t) = \begin{cases} \int\limits_0^t \frac{F_0}{T_0}\tau h(t-\tau)\mathrm{d}\tau & 0 < t \le T_0 \\ \int\limits_0^{T_0} \frac{F_0}{T_0}\tau h(t-\tau)\mathrm{d}\tau + \int\limits_{T_0}^{t}\left(-\frac{F_0}{T_0}\tau + 2F_0\right)h(t-\tau)\mathrm{d}\tau & T_0 < t \le 2T_0 \\ \int\limits_0^{T_0} \frac{F_0}{T_0}\tau h(t-\tau)\mathrm{d}\tau + \int\limits_{T_0}^{2T_0}\left(-\frac{F_0}{T_0}\tau + 2F_0\right)h(t-\tau)\mathrm{d}\tau & t > 2T_0 \end{cases}$$

5.2.2 *Response to Arbitrary Excitation*

When a system is subjected to harmonic excitation, its response is a harmonic vibration of the same frequency. When subjected to non-harmonic periodic excitation, the periodic excitation can be expanded into a Fourier series, and each term of the series can be treated as a harmonic excitation to determine its steady-state response. By summing up all harmonic steady-state responses based on the superposition principle of linear systems, the steady-state response of the system to the periodic excitation is obtained. For systems subjected to non-periodic, arbitrary time functions, or impact effects over an extremely short period, the problem is to find the response to any excitation. The classic solution to this problem is to first calculate the response of the vibration system to unit pulse excitation and then use the Duhamel's integral method to solve it.

In general, the n-th order differential equation for arbitrary excitation $f(t)$ is given by Eq. (5.23).

$$\frac{\mathrm{d}^n x}{\mathrm{d}t^n} + \alpha_{n-1}\frac{\mathrm{d}^{n-1} x}{\mathrm{d}t^{n-1}} + \cdots + \alpha_1 \frac{\mathrm{d}x}{\mathrm{d}t} + \alpha_0 x = f(t) \tag{5.23}$$

with the initial conditions

$$x(0) = 0, \quad \frac{\mathrm{d}x}{\mathrm{d}t}(0) = x_0^{(1)}, \ldots, \quad \frac{\mathrm{d}^{n-2} x}{\mathrm{d}t^{n-2}}(0) = x_0^{(n-2)}, \quad \frac{\mathrm{d}^{n-1} x}{\mathrm{d}t^{n-1}}(0) = x_0^{(n-1)}$$

A SDOF system under harmonic loads described by equation is $m\ddot{x}+c\dot{x}+kx = \mathrm{e}^{\mathrm{i}\lambda t}$. When $t \to \infty$, $\lim_{t\to\infty} x(t) = H\mathrm{e}^{\mathrm{i}\lambda t}$ is a harmonic function, H is called the frequency response function (FRF).

$$H(m, c, k, \lambda) = \frac{1}{-m\lambda^2 + c\mathrm{i}\lambda + k} \tag{5.24}$$

$$H = \frac{1/m}{(\omega^2 - \lambda^2) + \mathrm{i}2\eta\omega\lambda} \tag{5.25}$$

The relationship between the impulse response function (IRF) and the frequency response function is

$$x(t) = \frac{1}{2\pi}\int_{-\infty}^{\infty} X(\omega)\mathrm{e}^{\mathrm{i}\omega t}\mathrm{d}\omega$$

$$f(t) = \frac{1}{2\pi}\int_{-\infty}^{\infty} F(\omega)\mathrm{e}^{\mathrm{i}\omega t}\mathrm{d}\omega$$

$$X(\omega) = \int_{-\infty}^{\infty} x(t)\mathrm{e}^{-\mathrm{i}\omega t}\mathrm{d}t$$

$$F(\omega) = \int_{-\infty}^{\infty} f(t)\mathrm{e}^{-\mathrm{i}\omega t}\mathrm{d}t$$

$$\begin{aligned} x(t) &= \int_{0}^{t} h(t-\tau)f(\tau)\mathrm{d}\tau = \int_{-\infty}^{t} h(t-\tau)f(\tau)\mathrm{d}\tau \quad (f(t) = 0, \quad \forall t < 0) \\ &= \int_{-\infty}^{\infty} h(t-\tau)f(\tau)\mathrm{d}\tau \quad (h(t) = 0, \quad \forall t < 0) \end{aligned}$$

Then, we get

$$\begin{aligned}
x(t) &= \int_{-\infty}^{\infty} h(t-\tau)f(\tau)\mathrm{d}\tau \\
&= \int_{-\infty}^{\infty} h(t-\tau)\left[\frac{1}{2\pi}\int_{-\infty}^{\infty} F(\omega)\mathrm{e}^{\mathrm{i}\omega\tau}\mathrm{d}\omega\right]\mathrm{d}\tau \\
&= \frac{1}{2\pi}\int_{-\infty}^{\infty} F(\omega)\left[\int_{-\infty}^{\infty} h(t-\tau)\mathrm{e}^{\mathrm{i}\omega\tau}\mathrm{d}\tau\right]\mathrm{d}\omega \\
&= \frac{1}{2\pi}\int_{-\infty}^{\infty} F(\omega)\left[\int_{-\infty}^{\infty} h(u)\mathrm{e}^{\mathrm{i}\omega(t+u)}\mathrm{d}u\right]\mathrm{d}\omega \\
&= \frac{1}{2\pi}\int_{-\infty}^{\infty} F(\omega)H(\omega)\mathrm{e}^{\mathrm{i}\omega\tau}\mathrm{d}\omega \\
&\Rightarrow X(\omega) = F(\omega)H(\omega)
\end{aligned} \tag{5.26}$$

Convolution in the time domain is equivalent to multiplication in the frequency domain. This is one of the advantages of frequency domain analysis in linear vibration analysis.

$$h(t) * f(t) = \int_{0}^{t} h(t-\tau)f(\tau)\mathrm{d}\tau \Leftrightarrow H(\omega)F(\omega) \tag{5.27}$$

Consider

$$\ddot{x} + 2\eta\omega_n\dot{x} + \omega_n^2 x = \frac{f(t)}{m} \tag{5.28}$$

Introduce

$$x(t) = \frac{1}{2\pi}\int_{-\infty}^{\infty} X(\omega)\mathrm{e}^{\mathrm{i}\omega t}\mathrm{d}\omega$$

$$\dot{x}(t) = \frac{1}{2\pi}\int_{-\infty}^{\infty} \mathrm{i}\omega X(\omega)\mathrm{e}^{\mathrm{i}\omega t}\mathrm{d}\omega$$

$$\ddot{x}(t) = \frac{1}{2\pi} \int_{-\infty}^{\infty} -\omega^2 X(\omega) e^{i\omega t} d\omega$$

$$\Rightarrow \int_{-\infty}^{\infty} \left[\left(-\omega^2 + i2\eta\omega\omega_n + \omega_n^2\right) X(\omega) - \frac{F(\omega)}{m} \right] e^{i\omega t} d\omega = 0$$

$$\Rightarrow X(\omega) = \frac{F(\omega)/m}{\left(\omega_n^2 - \omega^2\right) + i2\eta\omega\omega_n} = F(\omega)H(\omega) \tag{5.29}$$

where, $H(\omega) = \frac{1/m}{(\omega_n^2 - \omega^2) + i2\eta\omega\omega_n}$.

5.3 SDOF System Under Random Excitations

Input–output relations for linear time-invariant (LTI) systems driven by random excitations $f(t)$ are discussed. $f(t)$ is a random process that can be completely specified and is not necessarily stationary or a Gaussian process. The equation is given by Eq. (5.18), with the initial conditions

$$x(0) = x_0, \quad \dot{x}(0) = \dot{x}_0$$

The response of Eq. (5.30) in the time domain is

$$x(t) = e^{-\eta\omega t} \left[x_0 \cos(\omega_d t) + \frac{x_0' + \eta\omega x_0}{\omega\sqrt{1 - \eta^2}} \sin(\omega_d t) \right] + \int_0^t h(t - \tau) f(\tau) d\tau \tag{5.30}$$

Given an ensemble of $f(t)$, the ensemble of $x(t)$ can be determined using this relation:

(1) Mean response

$$\langle x(t) \rangle = e^{-\eta\omega t} \left[x_0 \cos(\omega_d t) + \frac{\dot{x}_0 + \eta\omega x_0}{\omega\sqrt{1 - \eta^2}} \sin(\omega_d t) \right] + \left\langle \int_0^t h(t - \tau) f(\tau) d\tau \right\rangle$$
$$= e^{-\eta\omega t} \left[x_0 \cos(\omega_d t) + \frac{\dot{x}_0 + \eta\omega x_0}{\omega\sqrt{1 - \eta^2}} \sin(\omega_d t) \right] + \int_0^t h(t - \tau) \langle f(\tau) \rangle d\tau$$

$$m_x(t) = e^{-\eta\omega t} \left[x_0 \cos(\omega_d t) + \frac{\dot{x}_0 + \eta\omega x_0}{\omega\sqrt{1 - \eta^2}} \sin(\omega_d t) \right] + \int_0^t h(t - \tau) m_f(\tau) d\tau$$

(2) Autocorrelation function of $x(t)$ is derived from

$$\langle x(t_1)x(t_2)\rangle = \left\langle \int_0^{t_1}\int_0^{t_2} h(t_1-\tau_1)f(\tau_1)h(t_2-\tau_2)f(\tau_2)\mathrm{d}\tau_1\mathrm{d}\tau_2 \right\rangle$$

Get

$$\begin{aligned} R_{xx}(t_1,t_2) &= \left\langle \int_0^{t_1}\int_0^{t_2} h(t_1-\tau_1)f(\tau_1)h(t_2-\tau_2)f(\tau_2)\mathrm{d}\tau_1\mathrm{d}\tau_2 \right\rangle \\ &= \int_0^{t_1}\int_0^{t_2} h(t_1-\tau_1)h(t_2-\tau_2)R_{ff}(\tau_1,\tau_2)\mathrm{d}\tau_1\mathrm{d}\tau_2 \end{aligned} \tag{5.31}$$

(3) N-th order moment of the response process

Knowledge of the third order moment of the input is sufficient to determine the n-th order moment of the response process.

$$\begin{aligned} \langle x(t_1)x(t_2)x(t_3)\rangle &= \left\langle \int_0^{t_1}\int_0^{t_2}\int_0^{t_3} h(t_1-\tau_1)f(\tau_1)h(t_2-\tau_2)f(\tau_2)h(t_3-\tau_3)f(\tau_3)\mathrm{d}\tau_1\mathrm{d}\tau_2\mathrm{d}\tau_3 \right\rangle \\ &= \int_0^{t_1}\int_0^{t_2}\int_0^{t_3} h(t_1-\tau_1)h(t_2-\tau_2)h(t_3-\tau_3)\langle f(\tau_1)f(\tau_2)f(\tau_3)\rangle\mathrm{d}\tau_1\mathrm{d}\tau_2\mathrm{d}\tau_3 \end{aligned}$$

In general, for LTI systems, knowledge of the n-th order moment of the input is sufficient to determine the n-th order moment of the response process.

Note: this is not true for nonlinear systems.

5.3.1 SDOF System Under Gaussian White Noise Excitation

When the excitation is Gaussian white noise, the response features of the system, such as expectation, autocorrelation function, and variances will be discussed.

The system model and the initial conditions are the same as Eq. (5.18). The excitation $f(t)$ satisfies

$$\langle f(t)\rangle = 0, \quad \langle f(t_1)f(t_2)\rangle = I\delta(t_2-t_1)$$

The response $x(t)$ is

$$x(t) = \int_0^t h(t-\tau) f(\tau) \mathrm{d}\tau$$

$$\Rightarrow \langle x(t) \rangle = \int_0^t h(t-\tau) \langle f(\tau) \rangle \mathrm{d}\tau = 0$$

and

$$\langle x(t_1) x(t_2) \rangle = \left\langle \int_0^{t_1} \int_0^{t_2} h(t_1-\tau_1) f(\tau_1) h(t_2-\tau_2) f(\tau_2) \mathrm{d}\tau_1 \mathrm{d}\tau_2 \right\rangle$$

$$\begin{aligned} R_{xx}(t_1, t_2) &= \int_0^{t_1} \int_0^{t_2} h(t_1-\tau_1) h(t_2-\tau_2) \langle f(\tau_1) f(\tau_2) \rangle \mathrm{d}\tau_1 \mathrm{d}\tau_2 \\ &= \int_0^{t_1} \int_0^{t_2} h(t_1-\tau_1) h(t_2-\tau_2) I \delta(\tau_2-\tau_1) \mathrm{d}\tau_1 \mathrm{d}\tau_2 \\ &= \int_0^{t_1} \int_0^{t_2} I h(t_1-\tau_1) h(t_2-\tau_2) \mathrm{d}\tau_1 \mathrm{d}\tau_2 \end{aligned}$$

$$\begin{aligned} R_{xx}(t_1, t_2) &= \int_0^{t} I h(t_1-\tau) h(t_2-\tau) \mathrm{d}\tau \\ &= \int_0^{t_2} I \frac{1}{m\omega_\mathrm{d}} \exp[-\eta\omega(t_1-\tau)] \sin[\omega_\mathrm{d}(t_1-\tau)] \\ &\times \frac{1}{m\omega_\mathrm{d}} \exp[-\eta\omega(t_2-\tau)] \sin[\omega_\mathrm{d}(t_2-\tau)] \mathrm{d}\tau \\ &= \frac{I}{4\eta\omega^3 m^2} \exp[-\eta\omega(t_2-t_1)] \chi(t) \end{aligned} \tag{5.32}$$

where,

$$\begin{aligned} \chi(t) = &\left\{ \frac{\exp(-2\eta\omega t_1)}{1-\eta^2} \left[\eta^2 \cos\omega_\mathrm{d}(t_1+t_2) - \eta\sqrt{1-\eta^2} \sin\omega_\mathrm{d}(t_1+t_2) \right] \right. \\ &\left. - \cos\omega_\mathrm{d}(t_2-t_1) \right\} \\ &+ \left[\cos\omega_\mathrm{d}(t_2-t_1) + \frac{\eta}{\sqrt{1-\eta^2}} \sin\omega_\mathrm{d}(t_2-t_1) \right] \end{aligned}$$

When $t_1 = t_2 = t$, get

$$\begin{aligned}\sigma_x^2(t) &= R_{xx}(t, t) \\ &= \int_0^t I h^2(t-\tau)\mathrm{d}\tau \\ &= \frac{I}{4\eta\omega^3 m^2}\left\{\frac{\exp(-2\eta\omega t)}{1-\eta^2}\left[\eta^2\cos(2\omega_\mathrm{d}t) - \eta\sqrt{1-\eta^2}\sin(2\omega_\mathrm{d}t) - 1\right] + 1\right\}\end{aligned} \tag{5.33}$$

When $t \to \infty$ and $t_2 - t_1 = \tau$ is determined, has

$$\begin{aligned}&\lim_{\substack{t_1\to\infty \\ t_2\to\infty \\ t_2-t_1=\tau}} R_{xx}(t_1,\ t_2) \to \frac{I}{4\eta\omega^3 m^2}\exp(\eta\omega|\tau|)\left[\cos(\omega_\mathrm{d}\tau) + \frac{\eta}{\sqrt{1-\eta^2}}\sin(\omega_\mathrm{d}|\tau|)\right] \\ &\lim_{\substack{t_1\to\infty \\ t_2\to\infty \\ t_2-t_1=0}} R_{xx}(t_1, t_2) = \sigma_x^2 = \frac{I}{4\eta\omega^3 m^2}\end{aligned} \tag{5.34}$$

From Eq. (5.34), some conclusions can be drawn:

(1) For small times, the response is non-stationary:covariance is a function of t_1 and t_2, and variance is a function of time.
(2) For large times, the response becomes stationary: covariance is a function of time lags, and variance becomes time invariant.
(3) When the mean is 0, we say that the system reaches a stochastic steady state as time becomes large.
(4) If damping equals 0, the system fails to reach a steady state.

5.3.2 SDOF System Under Gaussian Modulated White Noise Excitation

When the white noise excitation is modulated by $e(t)$, Eq. (5.18) becomes

$$m\ddot{x} + c\dot{x} + kx = e(t)f(t) \tag{5.35}$$

The initial conditions are

$$x(0) = 0, \quad \dot{x}(0) = 0$$

and

$$\langle f(t)\rangle = 0, \quad \langle f(t_1)f(t_2)\rangle = I\delta(t_2 - t_1)$$

The modulated $e(t)$ is

$$e(t) = A[\exp(-\alpha t) - \exp(-\beta t)]$$

The response $x(t)$ is

$$x(t) = \int_0^t h(t-\tau)e(\tau)f(\tau)\mathrm{d}\tau$$

$$\Rightarrow \langle x(t)\rangle = \int_0^t h(t-\tau)e(\tau)\langle f(\tau)\rangle\mathrm{d}\tau = 0 \tag{5.36}$$

From

$$\langle x(t_1)x(t_2)\rangle = \left\langle \int_0^{t_1}\int_0^{t_2} h(t_1-\tau_1)e(\tau_1)f(\tau_1)h(t_2-\tau_2)e(\tau_2)f(\tau_2)\mathrm{d}\tau_1\mathrm{d}\tau_2 \right\rangle$$

The autocorrelation function in time t_1 and t_2, $t_2 > t_1$, is

$$\begin{aligned}
R_{xx}(t_1, t_2) &= \int_0^{t_1}\int_0^{t_2} h(t_1-\tau_1)h(t_2-\tau_2)e(\tau_1)e(\tau_2)\langle f(\tau_1)f(\tau_2)\rangle\mathrm{d}\tau_1\mathrm{d}\tau_2 \\
&= \int_0^{t_1}\int_0^{t_2} h(t_1-\tau_1)h(t_2-\tau_2)e(\tau_1)e(\tau_2)I\delta(\tau_2-\tau_1)\mathrm{d}\tau_1\mathrm{d}\tau_2 \\
&= \int_0^{t_2} Ih(t_1-\tau_2)h(t_2-\tau_2)e^2(\tau_2)\mathrm{d}\tau_2
\end{aligned} \tag{5.37}$$

$$\begin{aligned}
\sigma_x^2(t) &= \int_0^{t_1}\int_0^{t_2} h(t-\tau_1)h(t-\tau_2)e(\tau_1)e(\tau_2)I\delta(\tau_2-\tau_1)\mathrm{d}\tau_1\mathrm{d}\tau_2 \\
&= \int_0^{t} Ih^2(t-\tau)e^2(\tau)\mathrm{d}\tau
\end{aligned} \tag{5.38}$$

Let $f(t)$ be a stationary random process with zero mean, autocovariance is $C_{ff}(\tau)$, and its PSD is $S_{ff}(\omega)$.

By definition

$$S_{XX}(\omega) = \lim_{T\to\infty} \frac{1}{T}\left\langle |X_T(\omega)|^2 \right\rangle \tag{5.39}$$

Known

$$X_T(\omega) = H(\omega)F_T(\omega)$$

Therefore

$$S_{XX}(\omega) = \lim_{T\to\infty} \frac{1}{T}\left\langle H(\omega)F_T(\omega)F_T^*(\omega)H^*(\omega) \right\rangle$$

$$\Rightarrow S_{XX}(\omega) = \lim_{T\to\infty} \frac{1}{T}|H(\omega)|^2 S_{ff}(\omega). \tag{5.40}$$

5.4 Derivative and Integration of Random Processes

5.4.1 Definition of Derivative Random Processes

A process is said to have a derivative in the mean square sense if its second order moment process satisfies the condition:

$$\lim_{\Delta t\to 0} E\left\{[X(t+\Delta t) - X(t)]^2\right\} = 0 \tag{5.41}$$

This is often referred to as mean square continuity, which can be expressed as:

$$\lim_{\Delta t\to 0} X(t+\Delta t) = X(t)$$

The criteria for mean square continuity are:

(1) If the autocorrelation function $R_{xx}(t_1, t_2)$ is continuous in $t \in T(t_1 = t_2 = t)$, then the process is mean square continuous in $t \in T$.
(2) If the process is mean square continuous in $t \in T$, then $R_{xx}(t_1, t_2)$ is continuous in $t_1 = t_2 = t$.

$$\lim_{\Delta t\to 0} E\left\{\left[\frac{X(t+\Delta t) - X(t)}{\Delta t} - X'(t)\right]^2\right\} = 0 \tag{5.42}$$

or

$$\lim_{\Delta t\to 0} \frac{X(t+\Delta t) - X(t)}{\Delta t} = X'(t)$$

5.4.2 Mean Square Differentiability

To ascertain whether a process is mean square differentiable, one must solve for $X'(t)$ in Eq. (5.42).

The Cauchy criterion provides a means to test for mean square differentiability without explicitly determining $X(t)$. Specifically, if the process satisfies:

$$\lim_{\Delta t_1, \Delta t_2 \to 0} E\left\{\left[\frac{X(t+\Delta t_1) - X(t)}{\Delta t_1} - \frac{X(t+\Delta t_2) - X(t)}{\Delta t_2}\right]^2\right\} = 0 \tag{5.43}$$

$X(t)$ is considered differentiable in the mean square sense. If the mean and autocorrelation function of $\frac{\partial^2 R_{xx}(t_1,t_2)}{\partial t_1^2}$, $\frac{\partial^2 R_{xx}(t_1,t_2)}{\partial t_2^2}$, $\frac{\partial^2 R_{xx}(t_1,t_2)}{\partial t_1 \partial t_2}$ exist, Eq. (5.43) can be reformulated as

$$\begin{aligned}
&\lim_{\Delta t_1, \Delta t_2 \to 0} E\left\{\left[\frac{X(t+\Delta t_1) - X(t)}{\Delta t_1} - \frac{X(t+\Delta t_2) - X(t)}{\Delta t_2}\right]^2\right\} \\
&= \left[\frac{\partial^2 R_{xx}(t_1, t_2)}{\partial t_1^2} + \frac{\partial^2 R_{xx}(t_1, t_2)}{\partial t_2^2} - 2\frac{\partial^2 R_{xx}(t_1, t_2)}{\partial t_1 \partial t_2}\right]_{t_1=t_2=t} \\
&= 0
\end{aligned} \tag{5.44}$$

A sufficient and necessary condition for a random process to be mean square differentiable is the existence of all moments up to the second order.

5.4.3 Expectation and Autocorrelation Function of Derivatives

For a differentiable process $X(t)$ with derivative $Y(t)$, the expectation function is

$$\begin{aligned}
E[Y(t)] = E\left[\frac{\mathrm{d}X(t)}{\mathrm{d}t}\right] &= E\left[\lim_{\Delta t \to 0} \frac{X(t+\Delta t) - X(t)}{\Delta t}\right] \\
&= \lim_{\Delta t \to 0} E\left[\frac{X(t+\Delta t) - X(t)}{\Delta t}\right] \\
&= \lim_{\Delta t \to 0} \frac{E[X(t+\Delta t)] - E[X(t)]}{\Delta t} \\
&= \frac{\mathrm{d}E[X(t)]}{\mathrm{d}t}
\end{aligned}$$

Get

$$E\left[\frac{\mathrm{d}X(t)}{\mathrm{d}t}\right] = \frac{\mathrm{d}E[X(t)]}{\mathrm{d}t} \tag{5.45}$$

and the autocorrelation function is defined as

$$R_{yy}(t_1, t_2) = E[Y(t_1)Y(t_2)] = E\big[\dot{X}(t_1)\dot{X}(t_2)\big]. \tag{5.46}$$

5.4.4 Mean Square Integration of Random Processes

(1) Definition of Mean Square Integral

If the second order moment process $X(t)$ satisfies:

$$\lim_{\substack{\Delta t_1 \to 0,\\ n \to \infty}} E\left[\sum_{i=0}^{n-1} X(t_i')\Delta t_i - Y(t)\right]^2 = 0$$

The process $X(t)$ is said to be mean square integrable. And

$$Y = \lim_{\substack{\Delta t_i \to 0,\\ n \to \infty}} \sum_{i=0}^{n-1} X(t_i')\Delta t_i = \int_a^b X(t)\mathrm{d}t \tag{5.47}$$

is the mean square integral of process $X(t)$ in [a, b].

(2) The Condition of Mean Square Integrability

$$\int_a^b \int_a^b R_{xx}(t_1, t_2)\mathrm{d}t_1\mathrm{d}t_2 < \infty \tag{5.48}$$

5.4.5 Expectation and Autocorrelation Function of Random Process Integration

(1) The Expectation Function of Random Process Integration $Y(t)$

$$\begin{aligned} E[Y(t)] &= E\left[\int_a^b X(t)\mathrm{d}t\right] = E\left[\lim_{\substack{\Delta t_i \to 0,\\ n \to \infty}} \sum_{i=0}^{n-1} X(t_i')\Delta t_i\right] \\ &= \lim_{\substack{\Delta t_i \to 0,\\ n \to \infty}} \sum_{i=0}^{n-1} E[X(t_i)] \times \Delta t_i \end{aligned}$$

$$
\begin{aligned}
&= \int_a^b E[X(t)]\mathrm{d}t \\
&= \int_a^b m_X(t)\mathrm{d}t \\
&= m_Y
\end{aligned}
$$

$$
E\left[\int_a^b X(t)\mathrm{d}t\right] = \int_a^b E[X(t)]\mathrm{d}t \tag{5.49}
$$

It can be seen that the process of mean square integrability can exchange the order of integration and expectation from Eq. (5.49).

(2) Mean Square Value of Integration

$$
\begin{aligned}
E\left[Y^2\right] &= E\left[\int_a^b X(t_1)\mathrm{d}t_1 \times \int_a^b X(t_2)\mathrm{d}t_2\right] \\
&= E\left[\int_a^b \int_a^b X(t_1)X(t_2)\mathrm{d}t_1\mathrm{d}t_2\right] \\
&= \int_a^b \int_a^b E[X(t_1)X(t_2)]\mathrm{d}t_1\mathrm{d}t_2 \\
&= \int_a^b \int_a^b R_X(t_1, t_2)\mathrm{d}t_1\mathrm{d}t_2
\end{aligned} \tag{5.50}
$$

(3) Variance of Integration

$$
\begin{aligned}
D[Y(t)] &= E\left[Y^2(t)\right] - E^2[Y(t)] \\
&= \int_a^b \int_a^b R_{xx}(t_1, t_2)\mathrm{d}t_1\mathrm{d}t_2 - \int_a^b E[X(t_1)]\mathrm{d}t_1 \times \int_a^b E[X(t_2)]\mathrm{d}t_2 \\
&= \int_a^b \int_a^b \{R_{xx}(t_1, t_2) - E[X(t_1)] \cdot E[X(t_2)]\}\mathrm{d}t_1\mathrm{d}t_2 \\
&= \int_a^b \int_a^b C_{xx}(t_1, t_2)\mathrm{d}t_1\mathrm{d}t_2
\end{aligned} \tag{5.51}
$$

(4) Autocorrelation Function of Integration

Suppose $t_1, t_2 \in T$, and $Y(t) = \int_0^t X(\tau)\mathrm{d}\tau$ is random process. The autocorrelation function of $Y(t)$ is

$$\begin{aligned} R_{yy}(t_1, t_2) &= E[Y(t_1)Y(t_2)] \\ &= E\left[\int_0^{t_1} X(\tau)\mathrm{d}\tau \times \int_0^{t_2} X(v)\mathrm{d}v\right] \\ &= \int_0^{t_1}\int_0^{t_2} E\,[X(\tau)X(v)]\mathrm{d}v\mathrm{d}\tau \\ &= \int_0^{t_1}\int_0^{t_2} R_{xx}(\tau, v)\mathrm{d}v\mathrm{d}\tau. \end{aligned} \tag{5.52}$$

5.5 SDOF System Under Non-stationary Random Excitation

When a SDOF system is subjected to non-stationary random excitation, the input $f(t)$ in Eq. (5.19) is a random process. The inputs can be categorized into non-stationary deterministic excitations and stationary random excitations, as shown in Eq. (5.53).

$$f(t) = g(t)f_s(t) \tag{5.53}$$

If $g(t)$ is not constant, it represents a non-stationary process.
Taking the mean (expectation) of Eq. (5.18) yields, get

$$E[X(t)] = \int_0^t h(t-\tau)E[f(\tau)]\mathrm{d}\tau \tag{5.54}$$

If $E[f(\tau)] = 0$, $E[X(t)] = 0$, the autocorrelation function of the response is given by

$$\begin{aligned} R_{xx}(t_1, t_2) &= E[x(t_1)x(t_2)] \\ &= \int_0^{t_1}\int_0^{t_2} h(t_1-\tau_1)h(t_2-\tau_2)E\big[f(\tau_1)f(\tau_2)\big]\mathrm{d}\tau_1\mathrm{d}\tau_2 \end{aligned} \tag{5.55}$$

and

$$\begin{aligned} E[f(\tau_1)f(\tau_2)] &= g(\tau_1)g(\tau_2)E[f_s(\tau_1)f_s(\tau_2)] \\ &= g(\tau_1)g(\tau_2)R_{ff}(\tau_2 - \tau_1) \end{aligned} \tag{5.56}$$

Substituting Eq. (5.56) into Eq. (5.55) results in

$$R_{xx}(t_1, t_2) = \int_0^{t_1}\int_0^{t_2} h(t_1 - \tau_1)h(t_2 - \tau_2)g(\tau_1)g(\tau_2)R_{ff}(\tau_2 - \tau_1)\mathrm{d}\tau_1\mathrm{d}\tau_2 \tag{5.57}$$

Equation (5.57) implies that knowledge of the input autocorrelation function is required to compute the corresponding autocorrelation function of the response. If the auto-PSD of the input is denoted by $S_{ss}(\omega)$, then by the Wiener-Khinchine theorem, we have

$$R_{ss}(\tau) = \int_{-\infty}^{\infty} S_{ss}(\omega)\mathrm{e}^{\mathrm{i}\omega\tau}\mathrm{d}\omega$$

Insert into Eq. (5.57), get

$$R_{xx}(t_1, t_2) = \int_0^{t_1}\int_0^{t_2} h(t_1 - \tau_1)h(t_2 - \tau_2)g(\tau_1)g(\tau_2)\int_{-\infty}^{\infty} S_{ss}(\omega)\mathrm{e}^{\mathrm{i}\omega(\tau_2-\tau_1)}\mathrm{d}\omega\mathrm{d}\tau_1\mathrm{d}\tau_2$$

Organize again

$$R_{xx}(t_1, t_2) = \int_{-\infty}^{\infty} S_{ss}(\omega)\mathrm{e}^{\mathrm{i}\omega(\tau_2-\tau_1)}I(\omega, t_1)I^*(\omega, t_2)\mathrm{d}\omega \tag{5.58}$$

In Eq. (5.58),

$$I(\omega, t) = \int_0^{t} h(t - \tau)g(\tau)\mathrm{e}^{\mathrm{i}\omega\tau}\mathrm{d}\tau \tag{5.59}$$

$I(\omega, t)$ represents the response of a SDOF system to an input $g(\tau)\mathrm{e}^{\mathrm{i}\omega\tau}$ with initial conditions $x(0) = \dot{x}(0) = 0$. Since Eq. (5.58) is a function of frequency t_1 and t_2, and Eq. (5.59) is a function of time, the response $x(t)$ is a non-stationary process.

The subsequent sections will focus on calculating the statistics of the SDOF system under a single input.

(1) Autocorrelation Function Calculation

If the autocorrelation function of the input $f_s(t)$ is known, the response's autocorrelation function can be calculated using Eq. (5.57). If the auto-PSD function of the input $f_s(t)$ is known, it can be calculated using Eq. (5.59).

(2) Mean Square Calculation

Setting $t_1 = t_2 = t$ in Eqs. (5.57) and (5.58), the mean square of the response is given by

$$E\left[x^2(t)\right] = R_{xx}(t_1, t_2) = R_{xx}(t, t) \tag{5.60}$$

(3) Autocorrelation of Velocity Response

The autocorrelation of the velocity response $\dot{x}(t)$ can be derived from the derivative of the autocorrelation function:

$$R_{\dot{x}\dot{x}}(t_1, t_2) = \frac{\partial^2 R_{xx}(t_1, t_2)}{\partial t_1 \partial t_2} \tag{5.61}$$

(4) Cross-Correlation Calculation

The cross-correlation between $x(t)$ and $\dot{x}(t)$ is determined by

$$R_{x\dot{x}}(t_1, t_2) = \frac{\partial R_{xx}(t_1, t_2)}{\partial t_2} \tag{5.62}$$

(5) Frequency Domain Representation

Applying Fourier transform to $f(t)$ yields, get

$$F(\omega) = \frac{1}{2\pi} \int_{-\infty}^{\infty} f(t) \mathrm{e}^{-\mathrm{i}\omega t} \mathrm{d}t \tag{5.63}$$

$$f(t) = \int_{-\infty}^{\infty} F(\omega) \mathrm{e}^{\mathrm{i}\omega t} \mathrm{d}\omega \tag{5.64}$$

Equation (5.64) inserts into $x(t) = \int_0^t h(t-\tau) f(\tau) \mathrm{d}\tau$, get

$$x(t) = \int_0^t h(t-\tau) \int_{-\infty}^{\infty} F(\omega) e^{i\omega\tau} d\omega d\tau$$
$$= \int_{-\infty}^{\infty} F(\omega) \left[\int_0^t h(t-\tau) e^{i\omega\tau} d\tau \right] d\omega$$

Set $t - \tau = \theta$,

$$x(t) = \int_{-\infty}^{\infty} F(\omega) \left[-\int_t^0 h(\theta) e^{-i\omega\theta} d\theta \right] e^{i\omega t} d\omega$$
$$= \int_{-\infty}^{\infty} F(\omega) \left[\int_0^t h(\theta) e^{-i\omega\theta} d\theta \right] e^{i\omega t} d\omega$$

When θ locates out of $0 \le \theta \le t$, due to the integral in parentheses being equal to zero, therefore the upper limit of integration can expand from t to infinity and replace θ with τ, we get

$$x(t) = \int_{-\infty}^{\infty} F(\omega) \left[\int_0^t h(\tau) e^{-i\omega\tau} d\tau \right] e^{i\omega t} d\omega$$

Because $H(\omega) = \int_{-\infty}^{\infty} h(\tau) e^{-i\omega\tau} d\tau$, hence

$$x(t) = \int_{-\infty}^{\infty} F(\omega) H(\omega) e^{i\omega t} d\omega \tag{5.65}$$

(6) Mean Square Calculation in Frequency Domain

$$E[x(t)] = \int_{-\infty}^{\infty} H(\omega) E[F(\omega)] e^{i\omega t} d\omega \tag{5.66}$$

(7) Autocorrelation Function in Frequency Domain

$$R_{xx}(t_1, t_2) = E\left[x(t_1) x^*(t_2)\right]$$

$$= \int_{-\infty}^{\infty}\int_{-\infty}^{\infty} H(\omega_1)H^*(\omega_2)E\big[F(\omega_1)F^*(\omega_2)\big]e^{i(\omega_2 t_2-\omega_1 t_1)}d\omega_1 d\omega_2 \quad (5.67)$$

(8) Mean Square Value in Frequency Domain

$$E\big[x^2(t)\big] = \int_{-\infty}^{\infty}\int_{-\infty}^{\infty} H(\omega_1)H^*(\omega_2)E\big[F(\omega_1)F^*(\omega_2)\big]e^{i(\omega_2-\omega_1)t}d\omega_1 d\omega_2 \quad (5.68)$$

For the special case of a SDOF system subjected to a stationary stochastic process, setting $g(t) = 1$, so $f(t) = f_s(t)$ in Eq. (5.53), the correlation function and mean square value of the system's response can be derived from the corresponding formulas.

5.6 Random Vibration Under Random Initial Conditions

Considering the system's free vibration equation with initial conditions, the solution is given by:

$$x(t) = e^{-\xi\omega_0 t}[c_1\cos(\omega_d t) + c_2\sin(\omega_d t)] \quad (5.69)$$

where, $\omega_d = \omega_0\sqrt{1-\xi^2}$. When $t = 0$, $x = x_0$, $\dot{x} = \dot{x}_0$ $c_1 = x_0$, $c_2 = \frac{\dot{x}_0+\xi\omega_0 x_0}{\omega_d}$ therefore,

$$x(t) = e^{-\xi\omega_0 t}\left[\dot{x}_0,\ \cos(\omega_d t) + \frac{\dot{x}_0 + \xi\omega_0 x_0}{\omega_d}\sin(\omega_d t)\right]$$

or

$$x(t) = e^{-\xi\omega_0 t}\left\{x_0\left[\cos(\omega_d t) + \frac{\xi}{\sqrt{1-\xi^2}}\sin(\omega_d t)\right] + \frac{\dot{x}_0}{\omega_d}\sin(\omega_d t)\right\} \quad (5.70)$$

At the initial moment, where x_0 and $\dot{x}_0$ are random variables, the statistical values of $x(t)$ should be calculated using Eq. (5.70) as follows:

1. **Mean**

$$E[x(t)] = E(x_0)e^{-\xi\omega_0 t}\left[\cos(\omega_d t) + \frac{\xi}{\sqrt{1-\xi^2}}\sin(\omega_d t)\right] + E(\dot{x}_0)\frac{e^{-\xi\omega_0 t}\sin(\omega_d t)}{\omega_d}$$

or

$$m_x(t) = m_{x_0}\mathrm{e}^{-\xi\omega_0 t}\left[\cos(\omega_\mathrm{d}t) + \frac{\xi}{\sqrt{1-\xi^2}}\sin(\omega_\mathrm{d}t)\right] + m_{\dot{x}_0}\frac{\mathrm{e}^{-\xi\omega_0 t}\sin(\omega_\mathrm{d}t)}{\omega_\mathrm{d}} \tag{5.71}$$

2. **Autocorrelation Function**

$$\begin{aligned}
R_{xx}(t_1, t_2) &= E[x(t_1)x(t_2)] \\
&= E\left(x_0^2\right)\mathrm{e}^{-\xi\omega_0(t_2+t_1)}\left[\cos(\omega_\mathrm{d}t_1) + \frac{\xi}{\sqrt{1-\xi^2}}\sin(\omega_\mathrm{d}t_1)\right][\cos(\omega_\mathrm{d}t_2) \\
&+ \frac{\xi}{\sqrt{1-\xi^2}}\sin(\omega_\mathrm{d}t_2)\Bigg] \\
&+ E(x_0, \dot{x}_0)\mathrm{e}^{-\xi\omega_0(t_2+t_1)}\frac{1}{\omega_\mathrm{d}}\left\{\left[\cos(\omega_\mathrm{d}t_1) + \frac{\xi}{\sqrt{1-\xi^2}}\sin(\omega_\mathrm{d}t_1)\right]\sin(\omega_\mathrm{d}t_2)\right. \\
&\left.+\left[\cos(\omega_\mathrm{d}t_2) + \frac{\xi}{\sqrt{1-\xi^2}}\sin(\omega_\mathrm{d}t_2)\right]\sin(\omega_\mathrm{d}t_1)\right\} \\
&+ E\left(\dot{x}_0^2\right)\frac{\mathrm{e}^{-\xi\omega_0(t_2+t_1)}\sin(\omega_\mathrm{d}t)}{\omega_\mathrm{d}^2}\sin(\omega_\mathrm{d}t_1)\sin(\omega_\mathrm{d}t_2)
\end{aligned} \tag{5.72}$$

3. **Mean Square**

$$\begin{aligned}
E\left[x^2(t)\right] &= Ex_0^2\mathrm{e}^{-2\xi\omega_0}\left[\cos(\omega_\mathrm{d}t) + \frac{\xi}{\sqrt{1-\xi^2}}\sin(\omega_\mathrm{d}t)\right]^2 \\
&+ 2E(x_0, \dot{x}_0)\mathrm{e}^{-2\xi\omega_0 t}\frac{1}{\omega_\mathrm{d}}\left[\cos(\omega_\mathrm{d}t) + \frac{\xi}{\sqrt{1-\xi^2}}\sin(\omega_\mathrm{d}t)\right]\sin(\omega_\mathrm{d}t) \\
&+ E\dot{x}_0^2\frac{\mathrm{e}^{-2\xi\omega_0 t}}{\omega_\mathrm{d}^2}\sin^2(\omega_\mathrm{d}t)
\end{aligned} \tag{5.73}$$

Under random initial conditions, when the statistical properties of $E(x_0)$, $E(\dot{x}_0)$, $E(x_0^2)$, $E(\dot{x}_0^2)$ and $E(x_0, \dot{x}_0)$ are known, the statistical values of $\dot{x}(t)$ and $\ddot{x}(t)$ can be determined.

5.7 SDOF System Under Random Support Motions

Vibration systems are influenced not only by external loads but also by the motion or deformation of their supports, such as the vibration of building structures during earthquakes or the vibration caused by deforming road surfaces when vehicles travel. These vibrations originate from the support or support roughness.

5.7.1 Support Motion of a Structure

Figure 5.3 illustrates structural vibration due to support motion and its simplified representation. In Fig. 5.3, L is the height of the structure, m is the mass, E is the elastic modulus, I is the moment of inertia, $x(t)$ is the support motion, and $z(t)$ is the structural response. The stiffness is $k = 24EI/L^3$.

Figure 5.4 shows the modified structural vibration when using $z_t(t)$ which represents the total displacement of the structure.

Therefore, the kinematic equation is obtained when considering the support motion of a structure.

$$m\ddot{z}_t + c(\dot{z}_t - \dot{x}) + k(z_t - x) = 0 \tag{5.74}$$

Its initial conditions are

$$z_t(0) = z_{t0}, \quad \dot{z}_t(0) = \dot{z}_{t0}$$

From Eq. (5.74), we can deduce

$$\ddot{z}_i + 2\xi\omega_{\mathrm{n}}\dot{z}_t + \omega_{\mathrm{n}}^2 z_t = 2\xi\omega_{\mathrm{n}}\dot{x} + \omega_{\mathrm{n}}^2 x \tag{5.75}$$

where, ξ is the damper coefficient, ω_{n} is the natural frequency of the structure.

Set

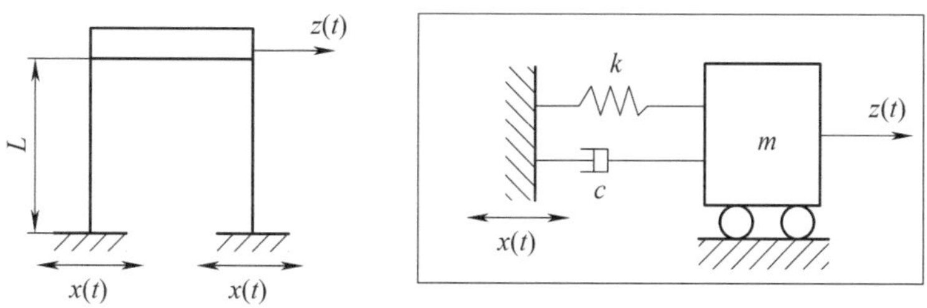

Fig. 5.3 The structural vibration from support motion

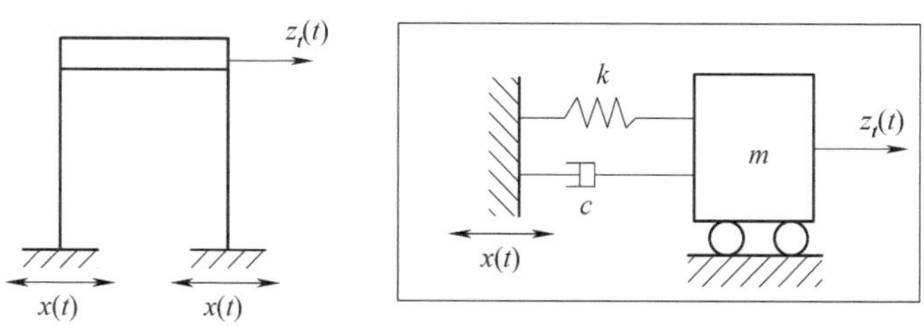

Fig. 5.4 The modified support of the structural vibration

$$z = z_t - x$$

where, z is the relative displacement, x is the support displacement.

Equation (5.75) changes to

$$m\ddot{z} + c\dot{z} + kz = -m\ddot{x} \tag{5.76}$$

and

$$z(0) = z_{t0} - x(0), \quad \dot{z}(0) = \dot{z}_{t0} - \dot{x}(0)$$

Equation (5.76) changes to

$$\ddot{z} + 2\eta\omega_n\dot{z} + \omega_n^2 z = -\ddot{x}. \tag{5.77}$$

5.7.2 *Support Roughness of a Vehicle*

Figure 5.5 illustrates vehicle traveling on rough road.

Based on Fig. 5.5, the kinematic equation can be constructed as follows:

$$m\ddot{u} + c\frac{\mathrm{d}}{\mathrm{d}t}[u - y(x)] + k[u - y(x)] = 0$$

With $u(0)$ and $\dot{u}(0)$ specified, we get

$$m\ddot{u} + c\frac{\mathrm{d}u}{\mathrm{d}t} + ku = c\frac{\mathrm{d}}{\mathrm{d}t}[y(x)] + k[y(x)].$$

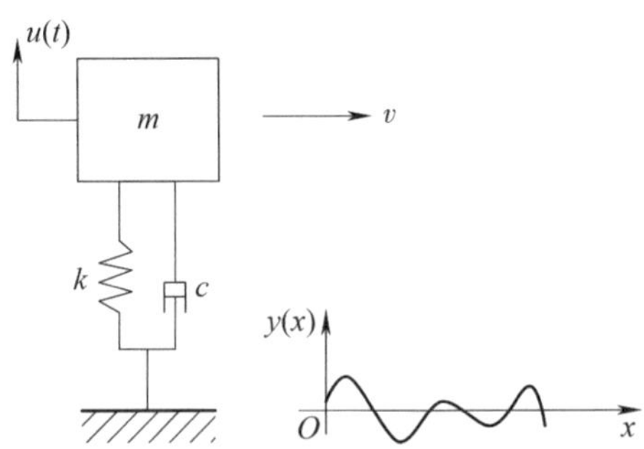

Fig. 5.5 Vehicle travelling on rough road

5.7.3 Steady State Random Vibration Analysis

1. **Analysis of Relative Displacement**

From Eq. (5.77) and its initial conditions

$$z(0) = z_{t0} - x(0), \quad \dot{z}(0) = \dot{z}_{t0} - \dot{x}(0)$$

The relative displacement is

$$z(t) = \mathrm{e}^{-\eta\omega t}[A\cos(\omega_{\mathrm{d}}t) + B\sin(\omega_{\mathrm{d}}t)] - \int_0^t h(t-\tau)m\ddot{x}(\tau)\mathrm{d}\tau \tag{5.78}$$

And in steady state

$$S_{zz}(\omega) = |H(\omega)|^2 m^2 S_{xx}(\omega) \tag{5.79}$$

$$\sigma_z^2 = \int_0^\infty |H(\omega)|^2 \omega^4 m^2 S_{xx}(\omega)\mathrm{d}\omega \tag{5.80}$$

2. **Doubly Supported SDOF System Under Differential Ground Motions**

In many cases, the motion of the support is asymmetric. For symmetric SDOF structures, the support stiffness and damping can be divided into equal halves, as shown in Fig. 5.6.

From Fig. 5.6, the following kinematic equations can be derived.

$$m\ddot{z}_t + \frac{c}{2}(\dot{z}_t - \dot{x}) + \frac{c}{2}(\dot{z}_t - \dot{y}) + \frac{k}{2}(z_t - x) + \frac{k}{2}(z_t - y) = 0 \tag{5.81}$$

$$m\ddot{z}_t + c\left(\dot{z}_t - \frac{\dot{x}+\dot{y}}{2}\right) + k\left(z_t - \frac{x+y}{2}\right) = 0 \tag{5.82}$$

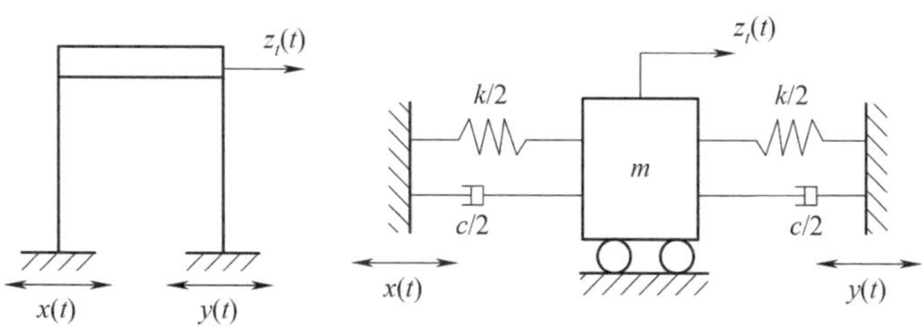

Fig. 5.6 The symmetric support motion

Total response includes pseudo-dynamic response and dynamic response in Eqs. (5.81) and (5.82). Equation (5.83) is the pseudo-dynamic response, and Eq. (5.84) is the dynamic response.

$$k\left(z_{\text{ps}} - \frac{x+y}{2}\right) = 0 \Rightarrow z_{\text{ps}} = \frac{x+y}{2} \tag{5.83}$$

$$z(t) = z_t(t) - z_{\text{ps}}(t) = z_t(t) - \frac{x+y}{2} \Rightarrow m\ddot{z} + c\dot{z} + kz = -m\frac{\ddot{x}+\ddot{y}}{2} \tag{5.84}$$

When the input is an aero mean, stationary, Gaussian random process with PSD matrix

$$\boldsymbol{S}(\omega) = \begin{pmatrix} S_{xx}(\omega) & S_{xy}(\omega) \\ S_{yx}(\omega) & S_{yy}(\omega) \end{pmatrix}$$

where

$$S_{xx}(\omega) = \lim_{T\to\infty}\frac{1}{T}\left\langle |x_T(\omega)|^2 \right\rangle$$

$$S_{xy}(\omega) = \lim_{T\to\infty}\frac{1}{T}\left\langle x_T(\omega)y_T^*(\omega) \right\rangle$$

$$S_{yx}(\omega) = \lim_{T\to\infty}\frac{1}{T}\left\langle x_T^*(\omega)y_T(\omega) \right\rangle = S_{xy}^*(\omega)$$

$$S_{xy}(\omega) = \left|S_{xy}(\omega)\right| \exp\left[-\mathrm{i}\varphi_{xy}(\omega)\right] = \left|S_{xy}(\omega)\right|\left[\cos\varphi_{xy}(\omega) - \mathrm{i}\sin\varphi_{xy}(\omega)\right]$$

$$S_{yx}(\omega) = S_{xy}^*(\omega) = \left|S_{xy}(\omega)\right|\left[\cos\varphi_{xy}(\omega) + \mathrm{i}\sin\varphi_{xy}(\omega)\right]$$

The force acting on the left spring is

$$\begin{aligned} F &= \frac{k}{2}(z_i - x) \\ &= \frac{k}{2}\left(z + \frac{x+y}{2} - x\right) \\ &= \frac{k}{4}[2z - (x-y)] \end{aligned}$$

Define

$$g(t) = \frac{4F}{k} = 2z - (x-y)$$

The following is to find the PSD of $g(t)$ and its variance.

From $\ddot{z} + 2\eta\omega_n\dot{z} + \omega_n^2 z = -\frac{\ddot{x}+\ddot{y}}{2}$, we can get

$$z_T(\omega) = -H_0(\omega)\frac{1}{2}\left[-\omega^2 x_T(\omega) - \omega^2 y_T(\omega)\right]$$
$$= H_0(\omega)\frac{\omega^2}{2}\left[x_T(\omega) + y_T(\omega)\right] \tag{5.85}$$

Known $H_0(\omega) = \frac{1}{(\omega_n^2-\omega^2)+i2\eta\omega\omega_n}$, therefore

$$S_{xx}(\omega) = \lim_{T\to\infty}\frac{1}{T}\left\langle |g_T(\omega)|^2\right\rangle$$

$$g_T(\omega) = 2z_T(\omega) - \left[x_T(\omega) + y_T(\omega)\right]$$

$$z_T(\omega) = H_0(\omega)\frac{\omega^2}{2}\left[x_T(\omega) + y_T(\omega)\right]$$

We can deduce

$$S_{gg}(\omega) = S_{xx}(\omega)H_1(\omega) + S_{yy}(\omega)H_2(\omega) + \left|S_{xy}(\omega)\right|H_3(\omega) \tag{5.86}$$

In Eq. (5.86), $\delta(t)$ is an artefact to remove singularity at $t = 0$ in the support displacement.

where,

$$H_1(\omega) = H_2(\omega) = \frac{1}{\omega^4} + \left\{\frac{1}{\left(\omega_n^2-\omega^2\right)^2 + (2\eta\omega\omega_n)^2} + \frac{2\left(\omega^2-\omega_n^2\right)}{\omega^2\left[\left(\omega_n^2-\omega^2\right)^2 + (2\eta\omega\omega_n)^2\right]}\right\}\left|H_f(\omega)\right|^2$$

$$H_3(\omega) = \left\{-\frac{2\cos(\varphi_{xy}(\omega))}{\omega^4} + \frac{2\cos[\varphi_{xy}(\omega)]}{(\omega_n^2-\omega^2)^2 + (2\eta\omega\omega_n)^2} + \frac{8\eta\omega\omega_n\sin(\varphi_{xy}(\omega))}{\omega^2\left[(\omega_n^2-\omega^2)^2 + (2\eta\omega\omega_n)^2\right]}\right\}\left|H_f(\omega)\right|^2$$

$H_f(\omega)$ is an artefact remove singularity at $\omega = 0$ in the support displacement.

$$\left|H_f(\omega)\right|^2 = \frac{\left(\omega/\omega_f\right)^4}{1-\left(\omega/\omega_f\right)^2 + 4\zeta_f^2\left(\omega/\omega_f\right)^2}$$

And

$$\sigma_g^2 = \int_0^{\infty} \left[S_{xx}(\omega)H_1(\omega) + S_{yy}(\omega)H_2(\omega) + \left|S_{xy}(\omega)\right|H_3(\omega)\right]\mathrm{d}\omega$$

The optimal solutions are produced neither by fully coherent nor by incoherent motions in Eqs. (5.81) and (5.82). From the pseudo-dynamic response and dynamic response, the results can be obtained.

$$\sigma_g^2 = \sigma_{\mathrm{ps}}^2 + \sigma_{\mathrm{d}}^2 + \sigma_{\mathrm{c}}^2 \tag{5.87}$$

where, σ_{ps}^2 contributes from pseudo-dynamic component, σ_{d}^2 contributes from dynamic component, and σ_{c}^2 contributes from correlation between pseudo-dynamic and dynamic component.

with

$$\sigma_{\mathrm{ps}}^2 = \int_0^{\infty} \left(\frac{1}{\omega^4}\left\{S_{xx}(\omega) + S_{yy}(\omega) - 2\cos\left[\varphi_{xy}(\omega)\right]\left|S_{xy}(\omega)\right|\right\}\right)\left|H_f(\omega)\right|^2\mathrm{d}\omega$$

$$\sigma_{\mathrm{d}}^2 = \int_0^{\infty} \left[\frac{1}{\left(\omega_{\mathrm{n}}^2 - \omega^2\right)^2 + (2\eta\omega\omega_{\mathrm{n}})^2}\right] \times \left\{S_{xx}(\omega) + S_{yy}(\omega) - 2\cos\left[\varphi_{xy}(\omega)\right]\left|S_{xy}(\omega)\right|\right\}\left|H_f(\omega)\right|^2\mathrm{d}\omega$$

$$\begin{aligned}\sigma_{\mathrm{c}}^2 &= \int_0^{\infty} \frac{1}{\omega^2\left[\left(\omega_{\mathrm{n}}^2 - \omega^2\right)^2 + (2\eta\omega\omega_{\mathrm{n}})^2\right]} \\ &= \left[\frac{1}{(\omega_{\mathrm{n}}^2 - \omega^2)^2 + (2)^2}\right]\left\{2(\omega^2 - \omega_{\mathrm{n}}^2)\left[S_{xx}(\omega) - S_{yy}(\omega)\right]\right. \\ &\quad \left. + 8\eta\omega\omega_{\mathrm{n}}\sin\left[\varphi_{xy}(\omega)\right]\left|S_{xy}(\omega)\right|\right\}\left|H_f(\omega)\right|^2\mathrm{d}\omega\end{aligned}$$

3. **Large Mass Concept**

Can we replace a given system with support motions by a modified equivalent system in which support displacements act as external forces? Figure 5.7 provides an answer to this question.

Suppose the support motion is $x(t) = \mathrm{e}^{\mathrm{i}\Omega t}$, Eq. (5.81) becomes

$$m\ddot{z}_t + c\dot{z}_t + kz_t = \frac{c}{2}\dot{x} + \frac{k}{2}x = \frac{1}{2}(\mathrm{i}\Omega c + k)\mathrm{e}^{\mathrm{i}\Omega t} \tag{5.88}$$

Get

$$\lim_{t\to\infty} z_t(t) \to H(\Omega)\mathrm{e}^{\mathrm{i}\Omega t}, \quad H(\Omega) = \frac{(\mathrm{i}\Omega c + k)/2}{-m\Omega^2 + \mathrm{i}\Omega c + k}$$

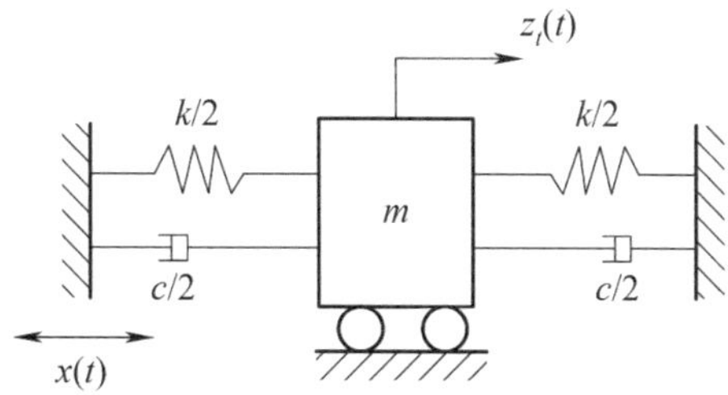

Fig. 5.7 The modified equivalent system in support displacements

From the equivalent system in support displacements, we replace the structure of the support motion $u(t)$ with a large mass, and the support motion is applied on the mass as an excitation. Figure 5.8 illustrates this concept.

From $x(t) = e^{i\Omega t}$ $\ddot{x}(t) = -\Omega^2 e^{i\Omega t}$ and Fig. 5.8, we get

$$\begin{cases} M\ddot{u} + \frac{c}{2}(\dot{u} - \dot{v}) + \frac{k}{2}(u - v) = M\Omega^2 e^{i\Omega t} \\ m\ddot{v} + \frac{c}{2}(\dot{v} - \dot{u}) + \frac{c}{2}\dot{v} + \frac{k}{2}(v - u) + \frac{k}{2}v = 0 \end{cases} \tag{5.89}$$

Equation (5.89) can be written as matrix:

$$\begin{pmatrix} M & 0 \\ 0 & m \end{pmatrix}\begin{pmatrix} \ddot{u} \\ \ddot{v} \end{pmatrix} + \frac{c}{2}\begin{pmatrix} 1 & -1 \\ -1 & 2 \end{pmatrix}\begin{pmatrix} \dot{u} \\ \dot{v} \end{pmatrix} + \frac{k}{2}\begin{pmatrix} 1 & -1 \\ -1 & 2 \end{pmatrix}\begin{pmatrix} u \\ v \end{pmatrix} = \begin{pmatrix} M\Omega^2 e^{i\Omega t} \\ 0 \end{pmatrix}$$

$$\begin{pmatrix} M & 0 \\ 0 & m \end{pmatrix}\begin{pmatrix} \ddot{u} \\ \ddot{v} \end{pmatrix} + c\begin{pmatrix} 0.5 & -0.5 \\ -0.5 & 1 \end{pmatrix}\begin{pmatrix} \dot{u} \\ \dot{v} \end{pmatrix} + k\begin{pmatrix} 0.5 & -0.5 \\ -0.5 & 1 \end{pmatrix}\begin{pmatrix} u \\ v \end{pmatrix} = \begin{pmatrix} M\Omega^2 e^{i\Omega t} \\ 0 \end{pmatrix}$$

When

$$\lim_{t\to\infty}\begin{pmatrix} u \\ v \end{pmatrix} = \begin{pmatrix} \boldsymbol{U} \\ \boldsymbol{V} \end{pmatrix} e^{i\Omega t}$$

can deduce

$$\begin{pmatrix} \boldsymbol{U} \\ \boldsymbol{V} \end{pmatrix} = \left[-\Omega^2\begin{pmatrix} M & 0 \\ 0 & m \end{pmatrix} + i\Omega c\begin{pmatrix} 0.5 & -0.5 \\ -0.5 & 1 \end{pmatrix} + k\begin{pmatrix} 0.5 & -0.5 \\ -0.5 & 1 \end{pmatrix}\right]^{-1}\begin{pmatrix} M\Omega^2 \\ 0 \end{pmatrix}$$

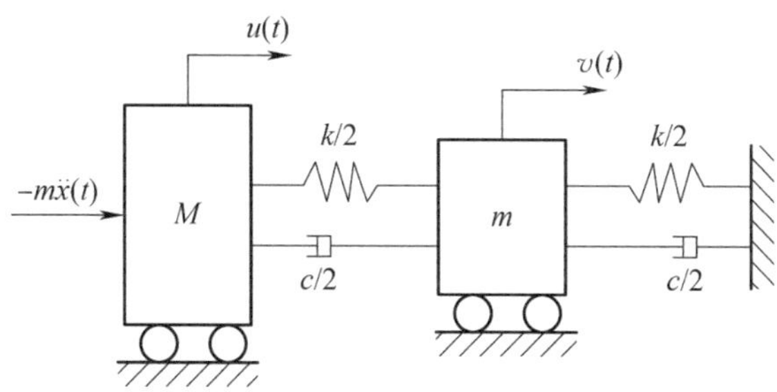

Fig. 5.8 The support motion of large mass and little mass

with

$$\lim_{M \to \infty} \boldsymbol{V}(\Omega) \to \boldsymbol{H}(\Omega)$$

After completing the above work, the previously learned knowledge of random vibration can be applied to study the statistical characteristics of related vibrations under random inputs.

Exercise 5

E5.1. Figure 5.9 shows a SDOF system diagram under the action of ideal white noise input. Calculate the autocorrelation function, spectral density function, and mean square value of the output. Known $m, k, c, S_x(\omega) = S_0$, find $R_y(\tau), S_y(\omega), \varphi_y^2$.

E5.2. Figure 5.10 illustrates a unicycle driving on uneven roads. Assuming the spatial power spectral function of the road surface is known, and the relation between spatial frequency and time frequency is $\Omega = \omega/(2\pi v)$.

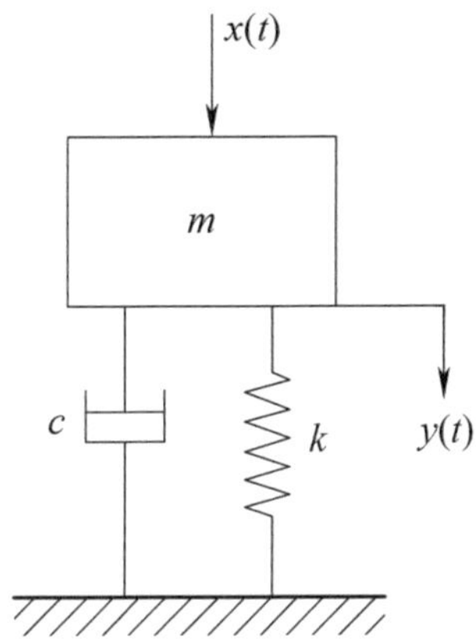

Fig. 5.9 The SDOF system diagram

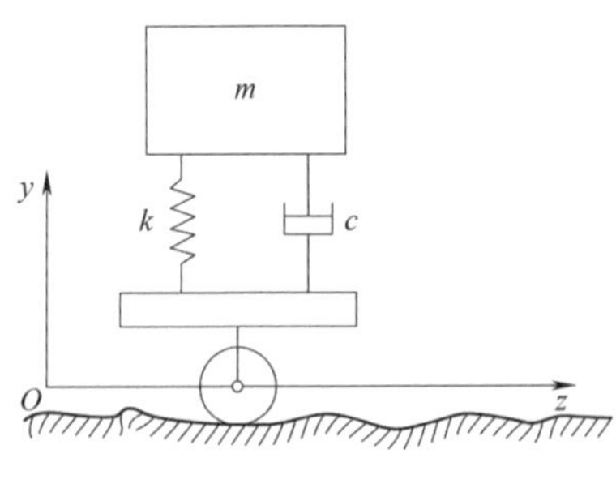

Fig. 5.10 The unicycle driving on uneven roads

When the unicycle travels at speed v, calculate the power spectral density, auto-correlation function, and mean of the displacement of the unicycle during vertical vibration. Known m, k, c, $S_y(\Omega) = S_0$, find $R_x(\tau)$, $S_x(\Omega)$, φ_x^2.

E5.3. On the SDOF system shown in Fig. 5.9, with Gaussian white noise with a spectral density of S_0, calculate the frequency at which the system's output response passes through a. Known m, k, c, $S_x(\omega) = S_0 a$, find r_0^+.

E5.4. Suppose the differential equation of a SDOF non-linear system is

$$m\ddot{x} + c_0\dot{x}^2 sgn\dot{x} + k_0 x = y(t)$$

Use the statistical linearization method to find the steady mean square responses of the system output excited by white noise.

Chapter 6
MDOF System

6.1 Preliminaries

Most engineering systems are complex and continuous, possessing an infinite number of degrees of freedom. Analyzing the vibration of such systems typically requires solving intricate partial differential equations, which often lack analytical solutions. In contrast, the analysis of multi-degree-of-freedom (MDOF) systems simplifies to solving a set of ordinary differential equations, offering a more tractable approach. Consequently, continuous systems are frequently approximated as MDOF systems to facilitate vibration analysis.

All concepts introduced in the previous chapter can be extended to MDOF systems. For instance, there is a unique equation of motion for each degree of freedom, and when generalized coordinates are employed, a corresponding generalized coordinate exists for each degree of freedom. These equations can be derived from Newton's second law or through influence coefficients as defined in Sect. 6.4. However, deriving the equations of motion using Lagrangian equation is often more convenient.

A system with n degrees of freedom has n natural frequencies, each associated with a distinct mode shape. The natural frequencies can be determined by solving the characteristic equation, which involves setting the determinant to zero. As the number of degrees of freedom increases, solving the characteristic equation becomes increasingly complex. Nevertheless, mode shapes exhibit the property of orthogonality, which is crucial for solving undamped forced vibration problems through modal analysis. Similarly, the solution for forced vibration problems in systems with viscous damping can be efficiently obtained using the concept of proportional damping.

The equations for a general MDOF system are derived using Newton's second law. Due to the complex algebraic manipulations involved in solving the scalar form of these equations, matrix representation is often used for MDOF systems.

J. Zhang and H. Zhou, *Random Vibration and Spectrum Analysis*,
https://doi.org/10.1007/978-981-95-0191-5_6

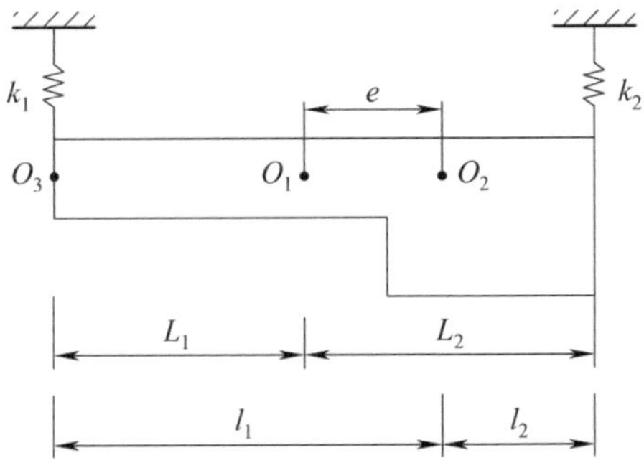

Fig. 6.1 The force analysis of a two DOF system

By expressing the set of n coupled equations in matrix form, the mass, damping, and stiffness matrices are identified.

Detailed derivations using influence coefficients, stiffness, flexibility, and inertia influence coefficients from first principles, as well as the expressions for potential and kinetic energies and their role in deriving the equations of motion based on Lagrange's equations, can be found in References.

Figure 6.1 presents an example of a two DOF system: a rigid bar supported by two springs. The two DOF consist of one translation and one rotation. k_1, k_2 are support stiffness. O_1 is elastic (graphic) center and O_2 is mass center. L_1, L_2 are O_1 to left and right support stiffness distance, separately. l_1, l_2 are O_2 to left and right support stiffness distance, separately.

From Fig. 6.1. get

$$\begin{aligned} &k_1L_1 = k_2L_2 \\ &m\ddot{y} + k_1(y - l_1\theta) + k_2(y + l_2\theta) = 0 \\ &I\ddot{\theta} - k_1(y - l_1\theta)l_1 + k_2(y + l_2\theta)l_2 = 0 \end{aligned}$$

In matrix form,

$$\begin{aligned} &\begin{pmatrix} m & 0 \\ 0 & I \end{pmatrix}\begin{pmatrix} \ddot{y} \\ \ddot{\theta} \end{pmatrix} + \begin{pmatrix} k_1 + k_2 & -k_1l_1 + k_2l_2 \\ -k_1l_1 + k_2l_2 & k_1l_1^2 + k_2l_2^2 \end{pmatrix}\begin{pmatrix} y \\ \theta \end{pmatrix} = \mathbf{0} \\ &\boldsymbol{M} = \begin{pmatrix} m & 0 \\ 0 & I \end{pmatrix}, \quad \boldsymbol{K} = \begin{pmatrix} k_1 + k_2 & -k_1l_1 + k_2l_2 \\ -k_1l_1 + k_2l_2 & k_1l_1^2 + k_2l_2^2 \end{pmatrix} \end{aligned} \tag{6.1}$$

where, $\boldsymbol{M}$ is diagonal and $\boldsymbol{K}$ is non-diagonal.

Because the elastic center and the mass center aren't at the same point, there must be coupled when built the kinematic equations of the two DOF system. Figure 6.2 gives the force analysis when choose O_3 as a reference point in Fig. 6.2.

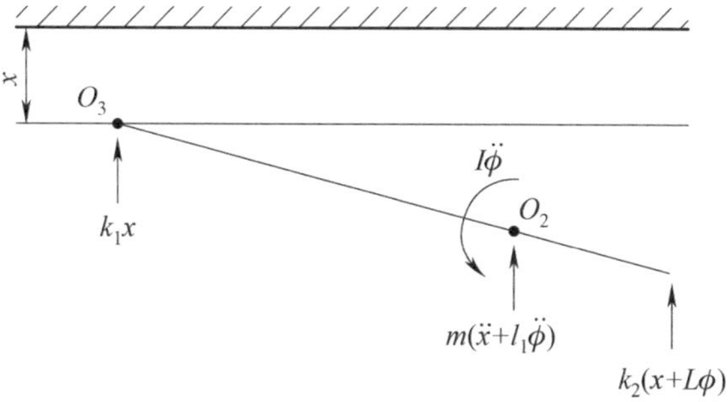

Fig. 6.2 The force analysis with reference O_3

$$\begin{aligned}&\begin{pmatrix} m & ml_1 \\ ml_1 & I \end{pmatrix}\begin{pmatrix} \ddot{x} \\ \ddot{\phi} \end{pmatrix} + \begin{pmatrix} k_1 + k_2 & k_2L \\ k_2L & k_2L^2 \end{pmatrix}\begin{pmatrix} x \\ \phi \end{pmatrix} = \mathbf{0} \\ &\boldsymbol{M} = \begin{pmatrix} m & ml_1 \\ ml_1 & I \end{pmatrix}, \quad \boldsymbol{K} = \begin{pmatrix} k_1 + k_2 & k_2L \\ k_2L & k_2L^2 \end{pmatrix}, \quad L = L_1 + L_2\end{aligned} \tag{6.2}$$

6.2 Uncoupled Equations of Motion

The MDOF system kinematic equation and its initial conditions,

$$\begin{aligned}&\boldsymbol{M}\ddot{\boldsymbol{X}} + \boldsymbol{C}\dot{\boldsymbol{X}} + \boldsymbol{K}\boldsymbol{X} = \boldsymbol{F}(t) \\ &\boldsymbol{X}(0) = \boldsymbol{X}_0, \quad \dot{\boldsymbol{X}}(0) = \dot{\boldsymbol{X}}_0\end{aligned} \tag{6.3}$$

Although the mass $\boldsymbol{M}$ and stiffness $\boldsymbol{K}$ matrices are generally non-diagonal, leading to coupled equations, we introduce a new set of dependent variables $\boldsymbol{Z}(t)$ through a transformation,

$$\boldsymbol{X}(t) = \boldsymbol{T}\boldsymbol{Z}(t) \tag{6.4}$$

where, $\boldsymbol{T}$ is a $n \times n$ transformation matrix, and the system in the new coordinate system becomes diagonal.

From Eq. (6.3), we can deduce

$$\begin{aligned}&\boldsymbol{M}\boldsymbol{T}\ddot{\boldsymbol{Z}}(t) + \boldsymbol{C}\boldsymbol{T}\dot{\boldsymbol{Z}}(t) + \boldsymbol{K}\boldsymbol{T}\boldsymbol{Z}(t) = \boldsymbol{F}(t) \\ &\boldsymbol{T}^{\mathrm{T}}\boldsymbol{M}\boldsymbol{T}\ddot{\boldsymbol{Z}}(t) + \boldsymbol{T}^{\mathrm{T}}\boldsymbol{C}\boldsymbol{T}\dot{\boldsymbol{Z}}(t) + \boldsymbol{T}^{\mathrm{T}}\boldsymbol{K}\boldsymbol{Z}(t) = \boldsymbol{T}^{\mathrm{T}}\boldsymbol{F}(t) \\ &\overline{\boldsymbol{M}}\ddot{\boldsymbol{Z}}(t) + \overline{\boldsymbol{C}}\dot{\boldsymbol{Z}}(t) + \overline{\boldsymbol{K}}\boldsymbol{Z}(t) = \overline{\boldsymbol{F}}(t)\end{aligned} \tag{6.5}$$

where, $\overline{\boldsymbol{M}}, \overline{\boldsymbol{C}}, \overline{\boldsymbol{K}}$ are the vibration system matrices in the new coordinate system, $\overline{\boldsymbol{F}}(t)$ is force vector in the new coordinate system.

The challenge lies in selecting the transformation matrix $\boldsymbol{T}$ such that the mass and stiffness matrices in the new coordinate system are diagonal. To achieve this, we consider the undamped free vibration analysis, seeking a special solution where all points on the system oscillate harmonically at the same frequency, as described by Eq. (6.6).

$$\boldsymbol{M}\ddot{\boldsymbol{X}} + \boldsymbol{K}\boldsymbol{X} = \boldsymbol{0} \tag{6.6}$$

Seek a special solution to this set of equations in which all points on the system oscillate harmonically at the same frequency.

$$x_1(t) = A_1 \mathrm{e}^{\mathrm{i}\omega t}$$
$$x_2(t) = A_2 \mathrm{e}^{\mathrm{i}\omega t}$$

That is

$$x_k(t) = r_k \mathrm{e}^{\mathrm{i}\omega t}, \quad k = 1, 2, \ldots, n$$

or

$$\boldsymbol{X}(t) = \boldsymbol{R}\mathrm{e}^{\mathrm{i}\omega t}$$

where, $\boldsymbol{R}$ is a $n \times 1$ vector, from it

$$\dot{\boldsymbol{X}}(t) = \mathrm{i}\omega R \mathrm{e}^{\mathrm{i}\omega t}$$
$$\ddot{\boldsymbol{X}}(t) = -\omega^2 R \mathrm{e}^{\mathrm{i}\omega t}$$

Procedure

$$\left(-\omega^2 \boldsymbol{M}\boldsymbol{R} + \boldsymbol{K}\boldsymbol{R}\right)\mathrm{e}^{\mathrm{i}\omega t} = \boldsymbol{0}$$
$$\Rightarrow -\omega^2 \boldsymbol{M}\boldsymbol{R} + \boldsymbol{K}\boldsymbol{R} = \boldsymbol{0}$$

hence

$$\boldsymbol{K}\boldsymbol{R} = \omega^2 \boldsymbol{M}\boldsymbol{R} \tag{6.7}$$

The algebraic eigenvalue problem arises from this process, with the mass matrix $\boldsymbol{M}$ being positive semi-definite and the stiffness matrix $\boldsymbol{K}$ being positive definite. The eigenvalue solutions are real-valued, and the eigenvalues are non-negative.

From Eq. (6.7), get

$$(\boldsymbol{K} - \omega^2 \boldsymbol{M})\boldsymbol{R} = \boldsymbol{0}$$

Let

$$(\boldsymbol{K} - \omega^2\boldsymbol{M})^{-1}(\boldsymbol{K} - \omega^2\boldsymbol{M})\boldsymbol{R} = \boldsymbol{0}$$

can deduce

$$\boldsymbol{IR} = \boldsymbol{0} \Rightarrow \boldsymbol{R} = \boldsymbol{0}$$

If $(\boldsymbol{K} - \omega^2\boldsymbol{M})^{-1}$ exists, $\boldsymbol{R} = \boldsymbol{0}$ is the solution of the Eq. (6.6). The condition for existence of nontrivial solution is that $(\boldsymbol{K} - \omega^2\boldsymbol{M})^{-1}$ does not exist. That is

$$\left|K - \omega^2 M\right| = 0 \tag{6.8}$$

The Eq. (6.8) is called the characteristic equation. This leads to the characteristic values $\omega_1^2 \leq \omega_2^2 \leq \cdots \omega_n^2$, and associated eigenvectors $\boldsymbol{R}_1, \boldsymbol{R}_2, \ldots, \boldsymbol{R}_n$.

6.3 Orthogonality Property of Eigenvectors

Eigenvectors are fundamental concepts in linear algebra with numerous practical applications, such as image processing, signal processing, and network analysis. They exhibit several key characteristics:

(1) Eigenvectors are non-zero vectors that correspond to an intrinsic value of a matrix, satisfying the equation $\boldsymbol{Ax} = \lambda\boldsymbol{x}$, where λ is an intrinsic value. If a vector meets this condition, it is an eigenvector of $\boldsymbol{A}$.
(2) Eigenvectors can represent the properties of matrices, such as symmetry, reversibility, and singularity. For example, if $\boldsymbol{A}$ is a symmetric matrix, its eigenvectors are orthogonal.
(3) Eigenvectors can decompose matrices; they can be used to express $\boldsymbol{A}$ as a linear combination of eigenvectors, which facilitates calculations like matrix inversion and power operations.
(4) Eigenvectors are instrumental in solving linear systems of equations. If A is a matrix and $\boldsymbol{b}$ is a vector, and if $\boldsymbol{A}$ has an eigenvector $\boldsymbol{x}$ such that $\boldsymbol{Ax} = \lambda\boldsymbol{x}$, then the linear equation system can be expressed in terms of a linear combination of eigenvectors.

In summary, eigenvectors are crucial in linear algebra and have a wide range of applications. This section focuses on discussing the orthogonal properties of eigenvectors.

Consider r- th and s-th eigenpairs, we get

$$\boldsymbol{KR}_r = \omega_r^2\boldsymbol{MR}_r$$

$$\boldsymbol{KR}_s = \omega_s^2\boldsymbol{MR}_s$$

$$\boldsymbol{R}_s^T\boldsymbol{KR}_r = \omega_r^2\boldsymbol{R}_s^T\boldsymbol{MR}_r \tag{6.9}$$

$$\boldsymbol{R}_r^T \boldsymbol{K} \boldsymbol{R}_s = \omega_s^2 \boldsymbol{R}_r^T \boldsymbol{M} \boldsymbol{R}_s \tag{6.10}$$

Transpose both sides of Eqs. (6.9) and (6.10), get

$$\boldsymbol{R}_s^{\mathrm{T}} \boldsymbol{K}^{\mathrm{T}} \boldsymbol{R}_r = \omega_s^2 \boldsymbol{R}_s^{\mathrm{T}} \boldsymbol{M}^{\mathrm{T}} \boldsymbol{R}_r$$

Since $\boldsymbol{K}^{\mathrm{T}} = \boldsymbol{K}, \quad \boldsymbol{M}^{\mathrm{T}} = \boldsymbol{M}$, we get

$$\boldsymbol{R}_s^{\mathrm{T}} \boldsymbol{K} \boldsymbol{R}_r = \omega_s^2 \boldsymbol{R}_s^{\mathrm{T}} \boldsymbol{M} \boldsymbol{R}_r \tag{6.11}$$

Substrate Eqs. (6.9) and (6.11),

$$\left(\omega_r^2 - \omega_s^2\right) \boldsymbol{R}_s^{\mathrm{T}} \boldsymbol{M} \boldsymbol{R}_r = \boldsymbol{0}$$

Introduce

$$\begin{aligned} \boldsymbol{\Phi} &= \begin{pmatrix} R_1 & R_2 & \cdots & R_n \end{pmatrix}_{(n \times n)} \\ \boldsymbol{\Lambda} &= diag\begin{pmatrix} \omega_1^2 & \omega_2^2 & \cdots & \omega_n^2 \end{pmatrix} \end{aligned}$$

Orthogonality relations are

$$\begin{aligned} \boldsymbol{\Phi}^{\mathrm{T}} \boldsymbol{M} \boldsymbol{\Phi} &= \boldsymbol{I} \\ \boldsymbol{\Phi}^{\mathrm{T}} \boldsymbol{K} \boldsymbol{\Phi} &= \boldsymbol{\Lambda} \end{aligned}$$

Select $\boldsymbol{T} = \boldsymbol{\Phi}$, consider undamped system

$$\boldsymbol{M}\ddot{\boldsymbol{X}} + \boldsymbol{K}\boldsymbol{X} = \boldsymbol{F}(t)$$

Initial conditions are

$$\boldsymbol{X}(0) = \boldsymbol{X}_0, \quad \dot{\boldsymbol{X}}(0) = \dot{\boldsymbol{X}}_0$$

From $\boldsymbol{X}(t) = \boldsymbol{\Phi}\boldsymbol{Z}(t)$, get

$$\begin{aligned} &\boldsymbol{M}\boldsymbol{\Phi}\ddot{\boldsymbol{Z}}(t) + \boldsymbol{K}\boldsymbol{\Phi}\boldsymbol{Z}(t) = \boldsymbol{F}(t) \\ &\boldsymbol{\Phi}^{\mathrm{T}}\boldsymbol{M}\boldsymbol{\Phi}\ddot{\boldsymbol{Z}}(t) + \boldsymbol{\Phi}^{\mathrm{T}}\boldsymbol{K}\boldsymbol{\Phi}\boldsymbol{Z}(t) = \boldsymbol{\Phi}^{\mathrm{T}}\boldsymbol{F}(t) \\ &\boldsymbol{I}\ddot{\boldsymbol{Z}}(t) + \boldsymbol{\Lambda}\boldsymbol{Z}(t) = \overline{\boldsymbol{F}}(t) \\ &\ddot{z}_r + \omega_r^2 z_r = f_r(t), \quad r = 1, 2, \ldots, n \end{aligned}$$

When the initial conditions become

$$X(0) = \Phi Z(0)$$
$$\boldsymbol{\Phi}^{\mathrm{T}} M X(0) = \boldsymbol{\Phi}^{\mathrm{T}} M \Phi Z(0) = Z(0)$$
$$Z(0) = \boldsymbol{\Phi}^{\mathrm{T}} M X(0), \dot{Z}(0) = \boldsymbol{\Phi}^{\mathrm{T}} M X'(0)$$

whereas

$$z_r(t) = z_r(0)\cos(\omega_r t) + \frac{\dot{z}_r(0)}{\omega_r}\sin(\omega_r t) + \int_0^t \frac{1}{\omega_r}\sin[\omega_r(t-\tau)]f_r(\tau)\mathrm{d}\tau$$

$$X(t) = \Phi Z(t)$$

$$x_k(t) = \sum_{r=1}^{n} \Phi_{kr} z_r(t)$$
$$= \sum_{r=1}^{n} \Phi_{kr}\left\{ z_r(0)\cos(\omega_r t) + \frac{\dot{z}_r(0)}{\omega_r}\sin(\omega_r t) + \int_0^t \frac{1}{\omega_r}\sin[\omega_r(t-\tau)]f_r(\tau)\mathrm{d}\tau \right\}$$

From $X(t) = TZ(t)$ and Eq. (6.3), we can get

$$\begin{aligned} &M\Phi\ddot{Z}(t) + C\Phi\dot{Z}(t) + K\Phi Z(t) = F(t) \\ &\boldsymbol{\Phi}^{\mathrm{T}} M\Phi\ddot{Z}(t) + \boldsymbol{\Phi}^{\mathrm{T}} C\Phi\dot{Z}(t) + \boldsymbol{\Phi}^{\mathrm{T}} K\Phi Z(t) = \boldsymbol{\Phi}^{\mathrm{T}} F(t) \end{aligned} \tag{6.12}$$

Equation (6.12) becomes

$$I\ddot{Z}(t) + \boldsymbol{\Phi}^{\mathrm{T}} C\Phi\dot{Z}(t) + \Lambda Z(t) = \overline{F}(t) \tag{6.13}$$

where, $\boldsymbol{\Phi} = (R_1 \quad R_2 \quad \cdots \quad R_n)_{(n\times n)}$ and $\boldsymbol{\Lambda} = diag(\omega_1^2 \quad \omega_2^2 \quad \cdots \quad \omega_n^2)$.

If $\boldsymbol{\Phi}^T C\boldsymbol{\Phi}$ is not a diagonal matrix, the equations of motion would still remain coupled.

6.4 Classical Damping Models

Damping is a fundamental material property that converts mechanical vibration energy into other forms of energy, typically thermal energy, through internal mechanisms. This property is also known as internal friction (at low frequencies) or ultrasonic attenuation (at high frequencies). Damping can be categorized into system damping, structural damping, and material damping. System damping involves installing specialized dampers, structural damping adds materials or structures to enhance a system's damping capacity, and material damping is inherent to the material itself. The damping discussed in this section refers to system damping.

If the damping matrix $\boldsymbol{C}$ is such that $\boldsymbol{\Phi}^{\mathrm{T}}\boldsymbol{C}\boldsymbol{\Phi}$, is a diagonal matrix, then the equations can be uncoupled, and $\boldsymbol{C}$ is referred to as a classical damping matrix. A well-known example is Rayleigh's proportional damping matrix.

$$C = \alpha \boldsymbol{M} + \beta \boldsymbol{K} \tag{6.14}$$

$\boldsymbol{C} = \boldsymbol{\Phi}^{T}\boldsymbol{C}\boldsymbol{\Phi}$ inserts Eq. (6.14), and $\boldsymbol{\Phi} = (R_1 \quad R_2 \quad \cdots \quad R_n)_{(n\times n)}$ and $\boldsymbol{\Lambda} = diag\left(\omega_1^2 \quad \omega_2^2 \quad \cdots \quad \omega_n^2\right)$, we get

$$\begin{aligned}
\boldsymbol{\Phi}^{\mathrm{T}}\boldsymbol{C}\boldsymbol{\Phi} &= \alpha\boldsymbol{\Phi}^{\mathrm{T}}\boldsymbol{M}\boldsymbol{\Phi} + \beta\boldsymbol{\Phi}^{\mathrm{T}}\boldsymbol{K}\boldsymbol{\Phi} = \alpha\boldsymbol{I} + \beta\boldsymbol{\Lambda} \\
\boldsymbol{\Phi}^{\mathrm{T}}\boldsymbol{C}\boldsymbol{\Phi} &= \boldsymbol{\Phi}^{\mathrm{T}}(\alpha\boldsymbol{M} + \beta\boldsymbol{K})\boldsymbol{\Phi} = \alpha\boldsymbol{I} + \beta diag(\omega_i^2) \\
c_n &= \alpha + \beta\omega_n^2
\end{aligned}$$

where

$$c_n = 2\eta_n\omega_n m_n, \quad \eta_n = \frac{\alpha}{2\omega_n} + \frac{\beta\omega_n}{2}$$

Equation (6.13) can be written as

$$\boldsymbol{I}\ddot{\boldsymbol{Z}}(t) + \boldsymbol{\Phi}^{\mathrm{T}}\boldsymbol{C}\boldsymbol{\Phi}\dot{\boldsymbol{Z}}(t) + \boldsymbol{\Lambda}\boldsymbol{Z}(t) = \overline{\boldsymbol{F}}(t)$$

and its initial conditions become

$$\boldsymbol{Z}(0) = \boldsymbol{\Phi}^{\mathrm{T}}\boldsymbol{M}\boldsymbol{X}(0), \quad \dot{\boldsymbol{Z}}(0) = \boldsymbol{\Phi}^{\mathrm{T}}\boldsymbol{M}\dot{\boldsymbol{X}}(0)$$

We can deduce

$$\ddot{z}_r + 2\eta_r\omega_r\dot{z}_r + \omega_r^2 z_r = f_r(t), \quad r = 1, 2, \ldots, n \tag{6.15}$$

With $z_r(0)$ and $\dot{z}_r(0)$ specified, get

$$\begin{aligned}
z_r(t) &= \exp(-\eta_r\omega_r t)[a_r\cos(\omega_{dr}t) + b_r\sin(\omega_{dr}t)] \\
&\quad + \int_0^t \frac{1}{\omega_{dr}} \exp[-\eta_r\omega_r(t-\tau)]f_r(\tau)\mathrm{d}\tau
\end{aligned} \tag{6.16}$$

From $\boldsymbol{X}(t) = \boldsymbol{\Phi}\boldsymbol{Z}(t)$, get

$$x_k(t) = \sum_{r=1}^{n} \Phi_{kr} z_r(t)$$

$$= \sum_{r=1}^{n} \Phi_{kr}\{\exp(-\eta_r\omega_r t)[a_r\cos(\omega_{dr}t) + b_r\sin(\omega_{dr}t)]$$

$$+ \int_0^t \frac{1}{\omega_{dr}} \exp[-\eta_r\omega_r(t-\tau)]f_r(\tau)\boldsymbol{d\tau}\} \quad (k = 1, 2, \ldots, n)$$

An example is given in Fig. 6.3 to illustrate the procedure of finding a structure's natural frequencies and its checking.

A structure of a three-layer slab supported by four supporting columns is shown in Fig. 6.3 shows, with masses of m_1, m_2, and m_3 for each slab floor. The support stiffness between the and foundation and the first floor is k_1, between the first and second floors is k_2, and between the second and third floors is k_3. The horizontal displacement of each layer is z_1, z_2 and z_3, respectively. The supporting column and slab are made of aluminum alloy, and the material characteristics and geometric dimensions are given in Table 6.1.

The equivalent model in Fig. 6.3 are shown in Fig. 6.4.

The characteristic parameters of the structure in Fig. 6.4 are

$$\begin{aligned}
&m_1\ddot{z}_1 + c_1(\dot{z}_1 - \dot{u}) + c_2(\dot{z}_1 - \dot{z}_2) + k_1(z_1 - u) + k_2(z_1 - z_2) = 0\\
&m_2\ddot{z}_2 + c_2(\dot{z}_2 - \dot{z}_1) + c_3(\dot{z}_2 - \dot{z}_3) + k_2(z_2 - z_1) + k_3(z_2 - z_3) = 0\\
&m_3\ddot{z}_3 + c_3(\dot{z}_3 - \dot{z}_2) + k_3(z_3 - z_2) = 0\\
&x_1 = z_1 - u\\
&x_2 = z_2 - u\\
&x_3 = z_3 - u
\end{aligned}$$

$$\begin{aligned}
&m_1\ddot{x}_1 + c_1\dot{x}_1 + c_2(\dot{x}_1 - \dot{x}_2) + k_1x_1 + k_2(x_1 - x_2) = -m_1\ddot{u}\\
&m_2\ddot{x}_2 + c_2(\dot{x}_2 - \dot{x}_1) + c_3(\dot{x}_2 - \dot{x}_3) + k_2(x_2 - x_1) + k_3(x_2 - x_3) = -m_2\ddot{u}\\
&m_3\ddot{x}_3 + c_3(\dot{x}_3 - \dot{x}_2) + k_3(x_3 - x_2) = -m_3\ddot{u}
\end{aligned}$$

Fig. 6.3 A three-layer structure

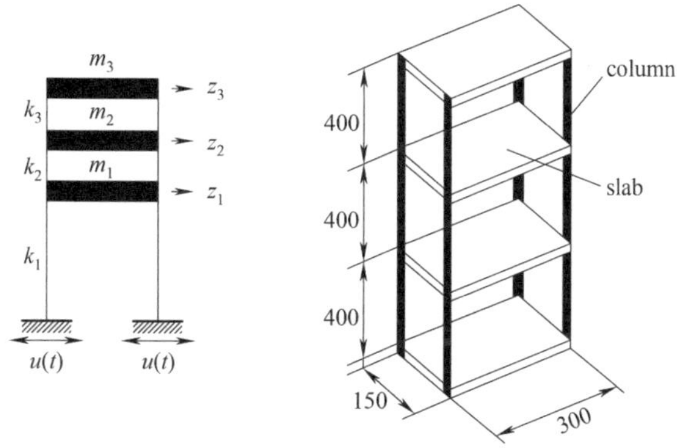

Table 6.1 The material characteristics and geometric dimensions

Part		Dimensions			
		Depth D		Width B	Length L
Column		$D_A = 3.00$		$B_A = 25.11$	$L_A = 400$
Slab		$D_B = 12.70$		$B_B = 150$	$L_B = 300$
Sq. No.	Part	Material	Mass/kg	Material properties	
				Young's modulus E/(N m^{-2}): 2.10×10^{11}	Mass density/(kg m^{-3}): 7890 kg/m^3
1	Column	Aluminum	0.081 4	6.9×10^{10}	2 700
2	Slab	Aluminum	1.543 0	6.9×10^{10}	2 700
3	Alleu screw M8	Steel	0.003 5		

Table 6.2 The parameters of Fig. 6.5

Sq. No.	Part	Qts Nos	Material	Mass/kg	Material properties		
					Young' s modulus E/ (N m^{-2})	Mass density/ (kg m^{-3})	Poisson' s ratio
1	Columns	3	Aluminum	$(m_1 + m_2 + m_3) =$ 0.326 4	6.9×10^{10}	2 700	0.3
2	Column	1	Steel	$m_4 = 0.303$ 7	2.0×10^{11}	7 800	0.3
3	Slab	1	Steel	$m_3 = 3.729$ 4	2.0×10^{11}	7 800	0.3

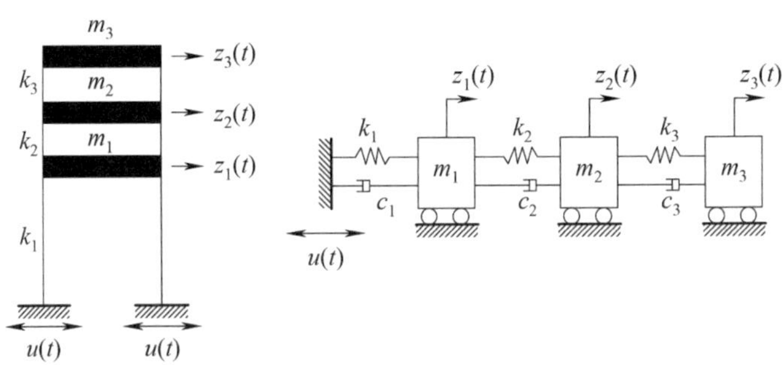

Fig. 6.4 A buliding frame under support motion

$$\begin{pmatrix} m_1 & 0 & 0 \\ 0 & m_2 & 0 \\ 0 & 0 & m_3 \end{pmatrix}\begin{pmatrix} \ddot{x}_1 \\ \ddot{x}_2 \\ \ddot{x}_3 \end{pmatrix} + \begin{pmatrix} c_1 + c_2 & -c_2 & 0 \\ -c_2 & c_2 + c_3 & -c_3 \\ 0 & -c_3 & c_3 \end{pmatrix}\begin{pmatrix} \dot{x}_1 \\ \dot{x}_2 \\ \dot{x}_3 \end{pmatrix}$$
$$+\begin{pmatrix} k_1 + k_2 & -k_2 & 0 \\ -k_2 & k_2 + k_3 & -k_3 \\ 0 & -k_3 & k_3 \end{pmatrix}\begin{pmatrix} x_1 \\ x_2 \\ x_3 \end{pmatrix} = \begin{pmatrix} m_1 \\ m_2 \\ m_3 \end{pmatrix} u$$

The structure's mass matrix (kg) is

$$\boldsymbol{M} = \begin{pmatrix} 1.896\,5 & 0 & 0 \\ 0 & 1.896\,5 & 0 \\ 0 & 0 & 1.733\,8 \end{pmatrix}$$

The structure's stiffness matrix (N/m) is

$$\boldsymbol{K} = 1 \times 10^4 \begin{pmatrix} 5.847\,5 & -2.923\,7 & 0 \\ -2.923\,7 & 5.847\,5 & -2.923\,7 \\ 0 & -2.923\,7 & 2.923\,7 \end{pmatrix}$$

And its damper factor vectors

$$\boldsymbol{\xi} = (0.024 \quad 0.009 \quad 0.007)^{\mathrm{T}}$$

The mass normalization modal matrix (kg) is

$$\boldsymbol{\Phi} = \begin{pmatrix} -0.246\,4 & 0.540\,1 & -0.418\,1 \\ -0.441\,6 & 0.213\,2 & 0.535\,6 \\ -0.545\,1 & -0.4560 & -0.267\,9 \end{pmatrix}$$

Get its natural frequencies (rad/s)

$$\boldsymbol{\omega}_{\mathrm{n}} = (17.893\,9 \quad 49.747\,6 \quad 71.119\,9)^{\mathrm{T}}$$

Orthogonality checks

$$\boldsymbol{\Phi}^{\mathrm{T}}\boldsymbol{M}\boldsymbol{\Phi} = \begin{pmatrix} 1 & 0 & 0 \\ 0 & 1 & 0 \\ 0 & 0 & 1 \end{pmatrix}$$

$$\boldsymbol{\Phi}^{\mathrm{T}}\boldsymbol{K}\boldsymbol{\Phi} = 1 \times 10^4 \times \begin{pmatrix} 0.320\,2 & 0 & 0 \\ 0 & 2.474\,8 & 0 \\ 0 & 0 & 5.058 \end{pmatrix}$$

Fig. 6.5 A sketch of a rectangular steel frame

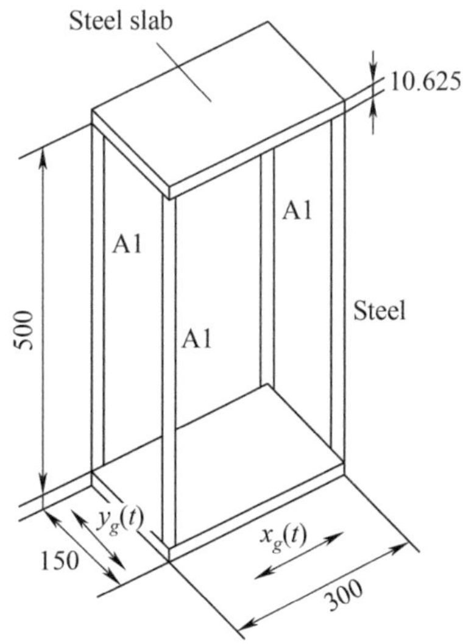

Another example is given in Fig. 6.5 to show on the procedure of finding a structure's natural frequencies and its checking in the different supports.

A structure of one laid slab supported by four supporting columns, in which three are Al and one is steel, as shown in Fig. 6.5. The diameter is 10.130 mm and 9.958 mm of the columns to aluminum and steel, separately.

We know the location of the mass center (m) of the structure is

$$\bar{x} = \frac{m_s\frac{b_s}{2} + (m_1 + m_4)\left(\frac{b_s - b_{s1}}{2}\right) + (m_2 + m_3)\left(\frac{b_s + b_{s1}}{2}\right)}{\rho_s t b_s d_s + m_1 + m_2 + m_3 + m_4}$$

$$\bar{y} = \frac{m_s\frac{d_s}{2} + (m_4 + m_3)\left(\frac{d_s - d_{s1}}{2}\right) + (m_1 + m_2)\left(\frac{d_s + d_{s1}}{2}\right)}{\rho_s t b_s d_s + m_1 + m_2 + m_3 + m_4}$$

$$m_1 = m_2 = m_3 = A_a\rho_a(h/2)$$

$$m_4 = A_s\rho_s(h/2), \quad m_s = \rho_s t b_s d_s$$

$$k_1 = k_2 = k_3 = \frac{12E_a I_a}{h^3}, \quad k_4 = \frac{12E_s I_s}{h^3}$$

$$m = m_{shb} + m_{\mathrm{colums}} = \rho_s t b_s d_s + 3\frac{A_s\rho_s h}{2} + \frac{A_s\rho_s h}{2}$$

$$\begin{aligned} I = {} & \frac{\rho_s t b_s d_s}{12}\left(b_s^2 + d_s^2\right) + \rho_s t b_s d_s\left[\left(\bar{x} - \frac{b_s}{2}\right)^2 + \left(\bar{y} - \frac{d_s}{2}\right)^2\right] \\ & + \frac{\rho_a A_a h}{2}\left(x_1^2 + y_1^2 + x_1^2 + y_1^2 + x_1^2 + y_1^2\right) \\ & + \frac{\rho_s A_s h}{2}\left(x_1^2 + y_1^2\right) + 3\frac{\rho_a A_a h}{2}\left(\frac{r_{a1}}{2}\right)^2 + \frac{\rho_s A_s h}{2}\left(\frac{r_s}{2}\right)^2 \end{aligned}$$

$$m\ddot{x} + k_1(x - y_1\theta - x_g) + c_1(\dot{x} - y_1\dot{\theta} - \dot{x}_g) + k_4(x - y_1\theta - x_g) + c_4(\dot{x} - y_1\dot{\theta} - \dot{x}_g)$$

$$+ k_8(x + y_2\theta - x_g) + c_8(\dot{x} + y_2\dot{\theta} - \dot{x}_g) + k_5(x + y_2\theta - x_g) + c_5(\dot{x} + y_2\dot{\theta} - \dot{x}_g) = 0$$

$$m\ddot{y} + k_2(y + x_1\theta - y_g) + c_2(\dot{y} + x_1\dot{\theta} - \dot{y}_g) + k_7(y + x_1\theta - y_g)$$
$$+ c_7(\dot{y} + x_1\dot{\theta} - \dot{y}_g) + k_3(y - x_2\theta - y_g) + c_3(\dot{y} - x_2\dot{\theta} - \dot{y}_g)$$
$$+ k_6(y - x_2\theta - y_g) + c_7(\dot{y} - x_2\dot{\theta} - \dot{y}_g) = 0$$

$$I\ddot{\theta} + x_1[k_2(y + x_1\theta - y_g) + c_2(\dot{y} + x_1\dot{\theta} - \dot{y}_g) + k_7(y + x_1\theta - y_g) + c_7(\dot{y} + x_1\dot{\theta} - \dot{y}_g)]$$
$$-x_2[k_3(y - x_2\theta - y_g) + c_3(\dot{y} - x_2\dot{\theta} - \dot{y}_g) + k_6(y - x_2\theta - y_g) + c_7(\dot{y} - x_2\dot{\theta} - \dot{y}_g)]$$
$$+y_2[k_8(x + y_2\theta - x_g) + c_8(\dot{x} + y_2\dot{\theta} - \dot{x}_g) + k_5(x + y_2\theta - x_g) + c_5(\dot{x} + y_2\dot{\theta} - \dot{x}_g)]$$
$$-y_1[k_1(x - y_1\theta - x_g) + c_1(\dot{x} - y_1\dot{\theta} - \dot{x}_g) + k_4(x - y_1\theta - x_g) + c_4(\dot{x} - y_1\dot{\theta} - \dot{x}_g)] = 0$$

$$u(t) = \begin{pmatrix} x(t) \\ y(t) \\ \theta(t) \end{pmatrix} \boldsymbol{M}\ddot{u} + \boldsymbol{C}\dot{u} + \boldsymbol{K}u = f(t)$$

$$bs = 0.30, \quad ds = 0.15, \quad ts = 10.625 \times 10^{-3}$$
$$bs_1 = 0.286, \quad ds_1 = 0.136$$
$$\overline{x} = 0.143\,6, \quad \overline{y} = 0.072\,0$$

the mass matrix is

$$\boldsymbol{M} = \begin{pmatrix} 4.044\,4(\text{kg}) & 0 & 0 \\ 0 & 4.044\,4(\text{kg}) & 0 \\ 0 & 0 & 0.043\,1(\text{kg m}^2) \end{pmatrix}$$

And the stiffness matrix (force in N, distance in m and angle in rad)

$$k_1 = k_2 = k_3 = k_4 = k_7 = k_8 = 3.424\,0 \times 10^3\ \text{N/m}$$
$$k_5 = k_6 = 9.267\,40 \times 10^3\ \text{N/m}$$
$$k^* = 313.129\,8\ \text{N/m}$$
$$K = 1 \times 10^4 \times \begin{pmatrix} 1.953\,9 & 0 & 0.047\,5 \\ 0 & 1.953\,9 & -0.082\,4 \\ 0.047\,5 & -0.082\,4 & 0.080\,5 \end{pmatrix}$$

and its damper factor vector is

$$\boldsymbol{\xi} = (0.02 \quad 0.009 \quad 0.007)^{\mathrm{T}}$$

Mass normalization modal matrix (kg) is

$$\boldsymbol{\Phi} = \begin{pmatrix} -0.245\,2 & 0.430\,8 & 0.039\,2 \\ 0.425\,4 & 0.248\,3 & -0.068\,1 \\ 0.761\,5 & 0 & 4.756\,7 \end{pmatrix}$$

We get the natural frequencies (rad/s)

$$\omega_n = (66.832\,1 \quad 69.506\,4 \quad 138.046\,0)^T$$

Orthogonality check is as follows

$$\boldsymbol{\Phi}^T \boldsymbol{M} \boldsymbol{\Phi} = \begin{pmatrix} 1 & 0 & 0 \\ 0 & 1 & 0 \\ 0 & 0 & 1 \end{pmatrix}$$

$$\boldsymbol{\Phi}^T \mathbf{K} \boldsymbol{\Phi} = 1 \times 10^4 \times \begin{pmatrix} 0.446\,7 & 0 & 0 \\ 0 & 0.483\,1 & 0 \\ 0 & 0 & 1.905\,7 \end{pmatrix}$$

From above two examples, we know that:

(1) Normal modes of vibration of a structure are special undamped free vibration solutions such that all points of the structure oscillate harmonically at the same frequency with the ratio of displacements at any two points being independent of time.
(2) Thus, for a structure vibration in one of its modes, the phase difference between oscillations at any two points is either 0 or π.
(3) The frequencies at which normal mode oscillations are possible are called the natural frequencies.
(4) (The modal matrix is orthogonal to the mass and stiffness matrices. This helps in diagonalising the mass and stiffness matrices.
(5) Undamped normal modes, in conjunction with proportional damping models, simplify vibration analysis procedures considerably.

6.5 Frequency Features of a Vibration System

Frequency features include spectrum analysis, frequency band energy analysis, and wavelet analysis. Spectrum analysis uses the Fourier transform to obtain the spectral distribution of vibration signals, revealing amplitude and peak values across different frequency bands. Frequency band energy analysis measures energy levels within various frequency bands to reflect the system's operational status.

In the time domain, the kinematic equation of a vibration system can be represented as Eq. (6.3). The Fourier transform pairs of a signal are well-known, and from these, we can deduce the complex frequency response matrix.

We have known Fourier transform pairs of a signal $x(t)$ are

$$X(\omega) = \int_{-\infty}^{\infty} x(t)\mathrm{e}^{-\mathrm{i}\omega t}\mathrm{d}t$$

$$x(t) = \frac{1}{2\pi}\int_{-\infty}^{\infty} X(\omega)\mathrm{e}^{\mathrm{i}\omega t}\mathrm{d}\omega$$

when $t \to \infty$

$$\boldsymbol{M}\left[\int_{-\infty}^{\infty} -\omega^2 \boldsymbol{X}(\omega)\mathrm{e}^{i\omega t}\boldsymbol{d}\omega\right] + C\left[\int_{-\infty}^{\infty} i\omega\boldsymbol{X}(\omega)\mathrm{e}^{i\omega t}\boldsymbol{d}\omega\right] + K\left[\int_{-\infty}^{\infty} \boldsymbol{X}(\omega)\mathrm{e}^{i\omega t}\boldsymbol{d}\omega\right]$$
$$= \int_{-\infty}^{\infty} X(\omega)\mathrm{e}^{i\omega t}\boldsymbol{d}\omega$$

We can deduce

$$\int_{-\infty}^{\infty} \left[\left(-\omega^2\boldsymbol{M} + \mathrm{i}\omega\boldsymbol{C} + \boldsymbol{K}\right)\boldsymbol{X}(\omega) - \boldsymbol{F}(\omega)\right]\mathrm{e}^{\mathrm{i}\omega t}\mathrm{d}\omega = \boldsymbol{0}$$
$$\boldsymbol{X}(\omega) = (-\omega^2\boldsymbol{M} + \boldsymbol{i}\omega\boldsymbol{C} + \boldsymbol{K})^{-1}\boldsymbol{F}(\omega)$$
$$\boldsymbol{X}(\omega) = \boldsymbol{H}(\omega)\boldsymbol{F}(\omega)$$

here

$$\boldsymbol{H}(\omega) = (-\omega^2\boldsymbol{M} + \mathrm{i}\omega\boldsymbol{C} + \boldsymbol{K})^{-1}$$

$\boldsymbol{H}(\omega)$ is the matrix of complex frequency response.

1. **MDOF System with s-th DOF Driven by a Unit Harmonic Force**

When a MDOF system is driven by a unit harmonic force at the s-th degree of freedom, as shown in Fig. 6.6, the response can be expressed as Eq. (6.17), with the kinematic equation format given by Eq. (6.18).

The response can be presented as

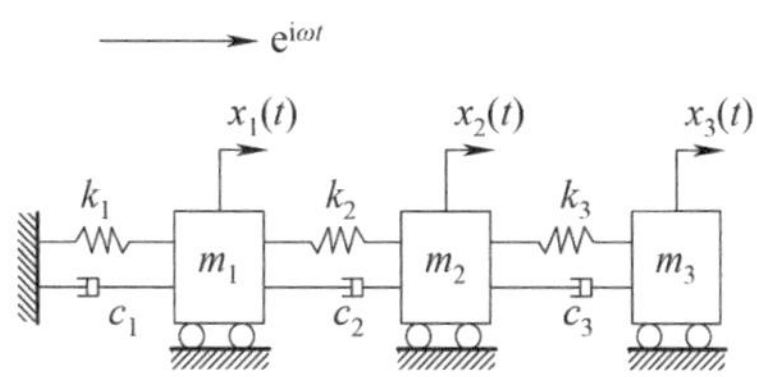

Fig. 6.6 A mechanical system with three masses

$$X_{rs}(t) = H_{rs}(t)\mathrm{e}^{i\omega t}$$
$$x_1(t) = X_{11}(t) = H_{11}(\omega)\mathrm{e}^{i\omega t}$$
$$x_2(t) = X_{12}(t) = H_{12}(\omega)\mathrm{e}^{i\omega t}$$
$$x_3(t) = X_{13}(t) = H_{13}(\omega)\mathrm{e}^{i\omega t}$$
$$\boldsymbol{H}(\omega) = \boldsymbol{H}_{rs}(\omega)$$

Its kinematic equation is

$$\boldsymbol{M}\ddot{\boldsymbol{X}} + \boldsymbol{C}\dot{\boldsymbol{X}} + \boldsymbol{K}\boldsymbol{X} = \boldsymbol{F}(t)\mathrm{e}^{\mathrm{i}\omega t} \tag{6.17}$$

$\boldsymbol{F}(t)$ format is

$$\boldsymbol{F}(t) = \begin{pmatrix} 0 \; 0 \cdots 1 \cdots 0 \; 0 \end{pmatrix} \tag{6.18}$$

Equation (6.18) means the s-th factor is 1 and others are 0.

The response of the r-th coordinate due to unit harmonic driving at the s-th coordinate is $X_{rs}(t)$, and the process for finding $X_{rs}(t)$ will be introduced subsequently.

$$\lim_{t\to\infty} \boldsymbol{X}(t) = \boldsymbol{X}_0\mathrm{e}^{i\omega t}$$

We can get

$$\dot{\boldsymbol{X}}(t) = \boldsymbol{X}_0\mathrm{i}\omega\mathrm{e}^{\mathrm{i}\omega t}$$

Continuously

$$-\boldsymbol{M}\boldsymbol{X}_0\omega^2\mathrm{e}^{i\omega t} + \boldsymbol{C}\boldsymbol{X}_0\mathrm{i}\omega\mathrm{e}^{i\omega t} + \boldsymbol{K}\boldsymbol{X}_0\mathrm{e}^{i\omega t} = \boldsymbol{F}\mathrm{e}^{i\omega t}$$
$$(-\omega^2\boldsymbol{M} + \mathrm{i}\omega\boldsymbol{C} + \boldsymbol{K})\boldsymbol{X}_0\mathrm{e}^{\mathrm{i}\omega t} = \boldsymbol{F}\mathrm{e}^{\mathrm{i}\omega t}$$
$$(-\omega^2\boldsymbol{M} + \mathrm{i}\omega\boldsymbol{C} + \boldsymbol{K})\boldsymbol{X}_0 = \boldsymbol{F}$$
$$\boldsymbol{X}(t) = \boldsymbol{X}_0\mathrm{e}^{\mathrm{i}\omega t} = \boldsymbol{\Phi}\boldsymbol{Z}_0\mathrm{e}^{\mathrm{i}\omega t}$$

Therefore

$$\boldsymbol{\Phi}^{\mathrm{T}}\boldsymbol{M}\boldsymbol{\Phi} = \boldsymbol{I}, \quad \boldsymbol{\Phi}^{\mathrm{T}}\boldsymbol{K}\boldsymbol{\Phi} = \boldsymbol{\Lambda}$$

If the damper matrix $\boldsymbol{C}$ is Rayleigh's proportional damping matrix, we can lead

$$\boldsymbol{\Phi}^{\mathrm{T}}\boldsymbol{M}\boldsymbol{\Phi} = \boldsymbol{\Gamma}$$

$\boldsymbol{\Gamma}$ is diagonal with

$$\Gamma_{mn} = 2\xi_n\omega_n$$

From

$$(-\omega^2\boldsymbol{M} + \mathrm{i}\omega\boldsymbol{C} + \boldsymbol{K})\boldsymbol{\Phi}\boldsymbol{Z}_0 = \boldsymbol{F} \tag{6.19}$$

Equation (6.19) times $\boldsymbol{\Phi}^{\mathrm{T}}$ both ends of equation, get

$$\begin{aligned}
&\boldsymbol{\Phi}^{\mathrm{T}}(-\omega^2\boldsymbol{M} + \mathrm{i}\omega\boldsymbol{C} + \boldsymbol{K})\boldsymbol{\Phi}Z_0 = \boldsymbol{\Phi}^{\mathrm{T}}\boldsymbol{F}\\
&(-\omega^2\boldsymbol{\Phi}^{\mathrm{T}}\boldsymbol{M}\boldsymbol{\Phi} + \mathrm{i}\omega\boldsymbol{\Phi}^{\mathrm{T}}\boldsymbol{C}\boldsymbol{\Phi} + \boldsymbol{\Phi}^{\mathrm{T}}\boldsymbol{K}\boldsymbol{\Phi})Z_0 = \boldsymbol{\Phi}^{\mathrm{T}}\boldsymbol{F}\\
&(-\omega^2\boldsymbol{I} + \mathrm{i}\omega\boldsymbol{\Gamma} + \boldsymbol{\Lambda})\boldsymbol{Z}_0 = \boldsymbol{\Phi}^{\mathrm{T}}\boldsymbol{F}
\end{aligned}$$

and set

$$Z_{0n} = \frac{\sum_{k=1}^{N}\Phi_{nk}^{t}F_k}{\omega_n^2 - \omega^2 + \mathrm{i}2\xi_n\omega_n\omega} \tag{6.20}$$

Because $\boldsymbol{F}(t) = (0\ 0\ \cdots\ 1\ \cdots\ 0\ 0)$, s-th entry equals 1 and the rests equal 0, Eq. (6.20) becomes

$$Z_{0n} = \frac{\Phi_{sn}}{\omega_n^2 - \omega^2 + \mathrm{i}2\xi_n\omega_n\omega}$$

From $\lim_{t\to\infty} X(t) = \Phi Z_0 \mathrm{e}^{i\omega t}$, we can deduce

$$X_r(t) = \sum_{n=1}^{N}\Phi_{rn}Z_{0n}\mathrm{e}^{\mathrm{i}\omega t} = \sum_{n=1}^{N}\frac{\Phi_{rn}\Phi_{sn}}{\omega_n^2 - \omega^2 + \mathrm{i}2\xi_n\omega_n\omega}\mathrm{e}^{\mathrm{i}\omega t} \tag{6.21}$$

and

$$\begin{aligned}
X_{rs}(t) &= H_{rs}(\omega)\mathrm{e}^{\mathrm{i}\omega t}\\
H_{rs}(\omega) &= \sum_{n=1}^{N}\frac{\Phi_{rn}\Phi_{sn}}{\omega_n^2 - \omega^2 + \mathrm{i}2\xi_n\omega_n\omega}
\end{aligned}$$

or

$$X_{rs}(t) = \sum_{n=1}^{N}\frac{\Phi_{rn}\Phi_{sn}}{\omega_n^2 - \omega^2 + \mathrm{i}2\eta_n\omega_n\omega}\mathrm{e}^{\mathrm{i}\omega t} \tag{6.22}$$

$$H_{rs}(\omega) = \sum_{n=1}^{N}\frac{\Phi_{rn}\Phi_{sn}}{\omega_n^2 - \omega^2 + \mathrm{i}2\xi_n\omega_n\omega} \tag{6.23}$$

For MDOF system, there are some features of response driven by an unit harmonic force in time and frequency domain, mainly in frequency. They are given as following

(1) $X_{rs}(t) = X_{sr}(t)$.
(2) $H_{rs}(\omega) = H_{sr}(\omega)$.
(3) $\boldsymbol{H}_{rs}(\omega) = \boldsymbol{H}_{sr}(\omega)$.
(4) $H(\omega)$ is symmetric but not Hermitian.
(5) $\boldsymbol{H}(\omega) = \left(-\omega^2\boldsymbol{M} + \mathrm{i}\omega\boldsymbol{C} + \boldsymbol{K}\right)^{-1} = \sum_{n=1}^{N} \frac{\Phi_{rn}\Phi_{sn}}{\omega_n^2-\omega^2+\mathrm{i}2\eta_n\omega_n\omega}$.
(6) $\boldsymbol{H}(\omega) = \left(-\omega^2\boldsymbol{M} + \mathrm{i}\omega\boldsymbol{C} + \boldsymbol{K}\right)^{-1}$, the equation is conceptually simple but it is computational difficult to implement.
(7) $\boldsymbol{H}(\omega) = \sum_{n=1}^{N} \frac{\Phi_{rn}\Phi_{sn}}{\omega_n^2-\omega^2+\mathrm{i}2\eta_n\omega_n\omega}$, the equation is computational easier to implement, but it is not all modes need to be included.

Example 6.1 A structure is given in Fig. 6.7. Find its eigenvalues, eigenvector and damper matrix. Known: $c_1 = c_2 = c_3 = 0, EI = 4.5 \times 10^6\ \mathrm{N\,m^3}$ for all columns and $\xi = 0.03$ for all modes.

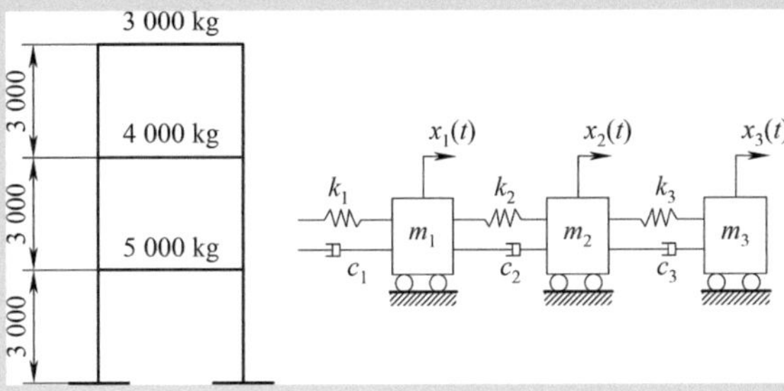

Fig. 6.7 A circuit with three blocks

Solution

From Fig. 6.7, the system kinematic equations are

$$m_1\ddot{x}_1 + k_1x_1 + k_2(x_1 - x_2) = 0$$
$$m_2\ddot{x}_2 + k_2(x_2 - x_1) + k_3(x_2 - x_3) = 0$$
$$m_3\ddot{x}_3 + k_3(x_3 - x_2) = 0$$

Rewrite by matrix format

$$\begin{pmatrix} 5\,000 & 0 & 0 \\ 0 & 4\,000 & 0 \\ 0 & 0 & 3\,000 \end{pmatrix}\begin{pmatrix} \ddot{x}_1 \\ \ddot{x}_2 \\ \ddot{x}_3 \end{pmatrix} + 4 \times 10^6 \begin{pmatrix} 2 & -1 & 0 \\ -1 & 2 & -1 \\ 0 & -1 & 1 \end{pmatrix}\begin{pmatrix} x_1 \\ x_2 \\ x_3 \end{pmatrix} = 0$$

$$\begin{vmatrix} 8 \times 10^6 - 5\,000\omega^2 & -4 \times 10^6 & 0 \\ -4 \times 10^6 & 8 \times 10^6 - 4\,000\omega^2 & -4 \times 10^6 \\ 0 & -4 \times 10^6 & 4 \times 10^6 - 3\,000\omega^2 \end{vmatrix} = 0$$

$$\omega^6 - 4\,933.3\omega^4 + 5.867 \times 10^7\omega^2 - 1.066 \times 10^9 = 0$$

The eigenvalues are

$$\boldsymbol{\omega} = (14.86 \quad 38.78 \quad 56.64)(\text{rad/s})$$

The eigenvector is

$$\boldsymbol{\varphi} = \begin{pmatrix} 0.005\,8 & 0.011\,4 & -0.006\,1 \\ 0.010\,0 & 0.001\,4 & 0.012\,2 \\ 0.012\,0 & -0.010\,7 & -0.008\,7 \end{pmatrix}$$

And the damper matrix is

$$\boldsymbol{C} = 1 \times 10^4 \times \begin{pmatrix} 1.140\,7 & -0.324\,4 & -0.065\,6 \\ -0.324\,4 & 0.954\,9 & -0.341\,9 \\ 0.065\,6 & -0.341\,9 & 0.584\,6 \end{pmatrix} (\text{N} \cdot \text{s} \cdot \text{m}^{-1})$$

2. MDOF System with *s*-th DOF Driven by a Unit Impulse Force

When a MDOF system is driven by a unit impulse force at the *s*-th degree of freedom, as depicted in Fig. 6.8. The response can be represented, and the initial conditions are known.

The damper matrix is Rayleigh's proportional damping matrix, and by applying Eq. (6.24) to both sides of the equation, we obtain Eq. (6.25).

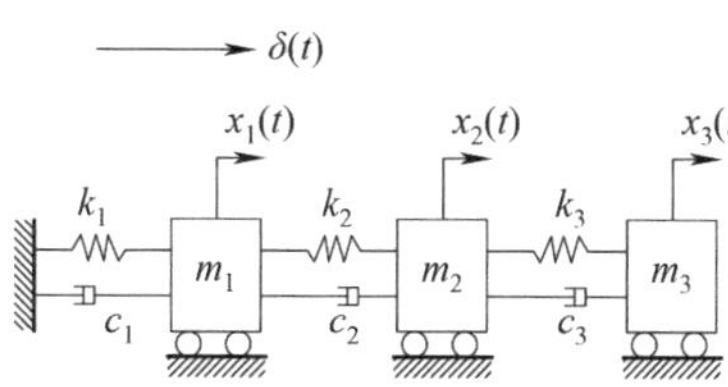

Fig. 6.8 A MDOF system is driven by a unit impulse

$$x_1(t) = X_{11}(t) = h_{11}(t)$$
$$x_2(t) = X_{12}(t) = h_{12}(t)$$
$$x_3(t) = X_{13}(t) = h_{13}(t)$$
$$\boldsymbol{h}(t) = \boldsymbol{h}_{rs}(t)$$

And known

$$F(\omega) = \int_{-\infty}^{\infty} f(t)\mathrm{e}^{-\mathrm{i}\omega t}\mathrm{d}t$$

$$f(t) = \frac{1}{2\pi}\int_{-\infty}^{\infty} F(\omega)\mathrm{e}^{\mathrm{i}\omega t}\mathrm{d}\omega$$

Its kinematic equations can be written as matrix format

$$\boldsymbol{M}\ddot{\boldsymbol{X}} + \boldsymbol{C}\dot{\boldsymbol{X}} + \boldsymbol{K}\boldsymbol{X} = \boldsymbol{F}\delta(t) \tag{6.24}$$

Initial conditions are

$$X(0) = X_0, \quad \dot{X}(0) = \dot{X}_0$$

The s-th entry, we can get

$$\boldsymbol{F} = \begin{pmatrix} 0 \; 0 \cdots 1 \cdots 0 \; 0 \end{pmatrix}$$

From $\boldsymbol{X}(t) = \boldsymbol{\Phi Z}(t)$, $\boldsymbol{\Phi}^{\mathrm{T}}\boldsymbol{M\Phi} = \boldsymbol{I}$, $\boldsymbol{\Phi}^{\mathrm{T}}\boldsymbol{K\Phi} = \boldsymbol{\Lambda}$, the damper matrix $\boldsymbol{C}$ is Rayleigh's proportional damping matrix, we can get

$$\boldsymbol{\Phi}^{\mathrm{T}}\boldsymbol{M\Phi} = \boldsymbol{\Gamma}$$

$\boldsymbol{\Gamma}$ is diagonal with

$$\Gamma_{nn} = 2\xi_n\omega_n$$

and

$$\boldsymbol{M\Phi}\ddot{\boldsymbol{Z}}(t) + \boldsymbol{C\Phi}\dot{\boldsymbol{Z}}(t) + \boldsymbol{K\Phi Z}(t) = \boldsymbol{F}\delta(t) \tag{6.25}$$

Equation (6.25) times $\boldsymbol{\Phi}^{\mathrm{T}}$ both ends of equation, get

$$\boldsymbol{\Phi}^{\mathrm{T}}\boldsymbol{M\Phi}\ddot{\boldsymbol{Z}}(t) + \boldsymbol{\Phi}^{\mathrm{T}}\boldsymbol{C\Phi}\dot{\boldsymbol{Z}}(t) + \boldsymbol{\Phi}^{\mathrm{T}}\boldsymbol{K\Phi Z}(t) = \boldsymbol{\Phi}^{\mathrm{T}}\boldsymbol{F}\delta(t) \tag{6.26}$$

Equation (6.26) can be written

$$\boldsymbol{I}\ddot{\boldsymbol{Z}} + \boldsymbol{\Gamma}\dot{\boldsymbol{Z}} + \boldsymbol{\Lambda}\boldsymbol{Z} = \boldsymbol{\Phi}^{\mathrm{T}}\boldsymbol{F}\delta(t)$$

Equation (6.24) becomes

$$\dot{z}_n + 2\xi_n\omega_n\dot{z}_n + \omega_n^2 z_n = \sum_{n=1}^{N} \Phi_{jn}F_j\delta(t) = \Phi_{ns}\delta(t) \tag{6.27}$$

Initial conditions are

$$z_n(0) = 0, \quad \dot{z}_n(0) = 0$$

get

$$z_n(t) = \Phi_{ns}h_n(t) = \frac{\Phi_{sn}}{\omega_{dn}} \exp(-\eta_n\omega_n t)\sin(\omega_{dn}t) \tag{6.28}$$

By $\boldsymbol{X}(t) = \boldsymbol{\Phi}\boldsymbol{Z}(t)$,

$$X_r(t) = \sum_{n=1}^{N} \Phi_{rn}z_n(t)$$

and

$$h_{rs}(t) = \sum_{n=1}^{N} \Phi_{rn}\Phi_{sn}\frac{1}{\omega_{dn}} \exp(-\eta_n\omega_n t)\sin(\omega_{dn}t)$$

To MDOF system, there have some features of response driven an unit impulse force. They are given as following.

(1) $h_{rs}(t) = h_{sr}(t)$.
(2) $\boldsymbol{h}(t) = \boldsymbol{h}_{sr}(t)$ is same as matrix of an impulse response function.
(3) $\boldsymbol{h}(t) = \boldsymbol{h}^{\mathrm{T}}(t)$.
(4) Not all modes need to be included in the summation.
(5) If an arbitrary load $f_s(\tau)$ is applied at the s-th DOF (instead of unit impulse excitation).

$$X_{rs}(t) = \int_0^t h_{sr}(t-\tau)f_s(\tau)\mathrm{d}\tau.$$

6.6 Dynamics of Euler–Bernoulli Beam

The Euler–Bernoulli beam theory is a fundamental equation in engineering mechanics and classical beam mechanics. It is a simplified linear elastic theory used to calculate the force and deformation characteristics of beams, as shown in Fig. 6.9.

The Euler–Bernoulli beam equation describes the relationship between the displacement and load of a beam:

$$\frac{\partial^2}{\partial x^2}\left(EI\frac{\partial^2 y}{\partial x^2}\right) = q \tag{6.29}$$

Among them: y is the displacement, it is the moment of the beam in brackets, and q is the distributed load on the beam.

1. **Euler–Bernoulli Beams Under Deterministic Excitation**

When the beam is subjected to a deterministic excitation, as shown in Fig. 6.9, Eq. (6.29) can be rewritten as Eq. (6.30), with boundary conditions(BCS) at the support point.

$$\frac{\mathrm{d}^2}{\mathrm{d}x^2}\left[EI(x)\frac{\mathrm{d}^2 y}{\mathrm{d}x^2}\right] = q(x) \tag{6.30}$$

Its boundary conditions at the support point are

$$y(0) = 0, \quad y(l) = 0, \quad \dot{y}(0) = 0, \quad EI\frac{\mathrm{d}^2 y}{\mathrm{d}x^2}(l) = 0$$

Consider $q(x)$ has related with the time we get

$$\frac{\partial^2}{\partial x^2}\left[EI(x)\frac{\partial^2 y}{\partial x^2} + a(x)\frac{\partial^3 y}{\partial x^2 \partial t}\right] + m(x)\frac{\partial^2 y}{\partial t^2} + c(x)\frac{\partial y}{\partial t} = q(x, t) \tag{6.31}$$

The boundary conditions become

$$y(0, t) = 0, \quad y(l, t) = 0, \quad \dot{y}(0, t) = 0, \quad EI\frac{\partial^2 y}{\partial x^2}(l, t) = 0$$

Initial conditions are

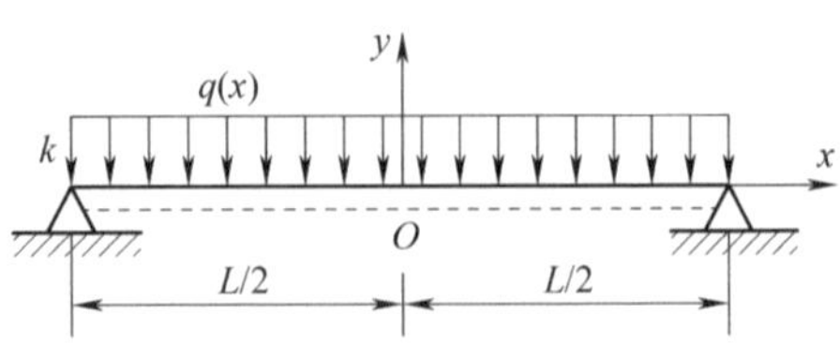

Fig. 6.9 A simply supported beam with a uniformly distributed load

$$y(x, 0) = y_0(x), \quad \frac{\partial y}{\partial t}(x, 0) = \dot{y}_0(x)$$

2. **Undamped Free Vibration Analysis**

When the beam's damper is neglected and there is no excitation, Eq. (6.31) becomes

$$\frac{\partial^2}{\partial x^2}\left[EI(x)\frac{\partial^2 y}{\partial x^2}\right] + m(x)\frac{\partial^2 y}{\partial t^2} = 0 \tag{6.32}$$

The boundary conditions are

$$y(0, t) = 0, \quad y(l, t) = 0, \quad \dot{y}(0, t) = 0, \quad EI\frac{\partial^2 y}{\partial x^2}(l, t) = 0$$

The response is

$$y(x, t) = \varphi(x)T(t)$$
$$[E\ddot{I}\ddot{\varphi}(x)]\ddot{T}(t) + \varphi(x)m(x)T(t) = 0$$

therefore,

$$[E\ddot{I}\ddot{\varphi}(x)]T(t) + \varphi(x)m(x)\ddot{T}(t) = 0$$

We get

$$\frac{[E\ddot{I}\ddot{\varphi}(x)]T(t)}{\varphi(x)T(t)} + \frac{\varphi(x)m(x)\ddot{T}(t)}{\varphi(x)T(t)} = 0$$
$$\frac{E\ddot{I}\ddot{\varphi}(x)}{\varphi(x)m(x)} = -\frac{\ddot{T}(t)}{T(t)} = \text{constant} = \omega^2$$

From $\ddot{T}(t) + \omega^2 T(t) = 0$, we get

$$T(t) = A\cos(\omega t) + b\sin(\omega t)$$

and

$$[E\ddot{I}\ddot{\varphi}(x)] - \omega^2 m(x)\varphi(x) = 0$$

Under the following boundary conditions

$$y(0, t) = 0, \quad y(l, t) = 0, \quad \dot{y}(0, t) = 0, \quad EI\frac{\partial^2 y}{\partial x^2}(l, t) = 0$$

We get

$$\varphi(0) = 0, \quad \varphi(t) = 0, \quad \dot{\varphi}(0) = 0, \quad EI\frac{\mathrm{d}^2\varphi}{\mathrm{d}x^2}(l) = 0$$

3. **Eigenvalue Problem**

From Eq. (6.32), we can deduce

$$[E\ddot{I}\ddot{\varphi}(x)] - \omega^2 m(x)\varphi(x) = 0 \tag{6.33}$$

known

$$\varphi(0) = 0, \quad \varphi(t) = 0, \quad \dot{\varphi}(0) = 0, \quad EI\frac{\mathrm{d}^2\varphi}{\mathrm{d}x^2}(l) = 0$$

When the section and material don't change with x, we get

$$EI(x) = EI, \quad m(x) = m$$

Equation (6.33) turns

$$EI\ddot{\varphi}(x) - \omega^2 m\varphi(x) = 0 \tag{6.34}$$

Set

$$\ddot{\varphi} - \lambda^4\varphi = 0 \tag{6.35}$$

where, $\lambda^4 = \frac{\omega^2 m}{EI}$, and the solution of Eq. (6.35) is

$$\varphi(x) = a[\cos(\lambda x) + \cosh(\lambda x)] + b[\cos(\lambda x) - \cosh(\lambda x)]$$

$$+c[\sin(\lambda x) + \sinh(\lambda x)] + d[\sin(\lambda x) - \sinh(\lambda x)]$$

Differential of $\varphi(x)$ is

$$\begin{aligned}
\dot{\varphi}(x) &= a\lambda[-\sin(\lambda x) + \sinh(\lambda x)] + b\lambda[-\sin(\lambda x) - \sinh(\lambda x)] \\
&+c\lambda[\cos(\lambda x) + \cosh(\lambda x)] + d\lambda[\cos(\lambda x) - \cosh(\lambda x)] \\
&= a\lambda^2[-\cos(\lambda x) + \cosh(\lambda x)] + b\lambda^2[-\cos(\lambda x) + \cosh(\lambda x)] \\
&+c\lambda^2[-\sin(\lambda x) + \sinh(\lambda x)] + d\lambda^2[-\sin(\lambda x) + \sinh(\lambda x)]
\end{aligned}$$

From the initial conditions, we can find

(1) From $\varphi(0) = 0$, get $a = 0$.
(2) From $\dot{\varphi}(0) = 0$, get $c = 0$.
(3) From $\varphi(l) = 0$, get $b[\cos(\lambda l) - \cosh(\lambda l)] + d[\sin(\lambda l) - \sinh(\lambda l)] = 0$.
(4) From $\ddot{\varphi}(l) = 0$, get $b\lambda^2[-\cos(\lambda l) - \cosh(\lambda l)] + d\lambda^2[-\sin(\lambda l) - \sinh(\lambda l)] = 0$.

(5) From $\varphi(l) = 0$, get $b[\cos(\lambda l) - \cosh(\lambda l)] + d[\sin(\lambda l) - \sinh(\lambda l)] = 0$.
(6) From $\ddot{\varphi}(l) = 0$, get $b\lambda^2[-\cos(\lambda l) - \cosh(\lambda l)] + d\lambda^2[-\sin(\lambda l) - \sinh(\lambda l)] = 0$

and

$$\begin{pmatrix} \cos(\lambda l) - \cosh(\lambda l) & \sin(\lambda l) - \sinh(\lambda l) \\ \lambda^2[-\cos(\lambda l) - \cosh(\lambda l)] & \lambda^2[-\sin(\lambda l) - \sinh(\lambda l)] \end{pmatrix} \begin{pmatrix} b \\ d \end{pmatrix} = \mathbf{0}$$

$$\begin{vmatrix} \cos(\lambda l) - \cosh(\lambda l) & \sin(\lambda l) - \sinh(\lambda l) \\ \lambda^2[-\cos(\lambda l) - \cosh(\lambda l)] & \lambda^2[-\sin(\lambda l) - \sinh(\lambda l)] \end{vmatrix} = 0$$

$$\varphi_n(x) = \cos(\lambda_n x) - \cosh(\lambda_n x) + \sigma_n[\sin(\lambda_n x) - \sinh(\lambda_n x)]$$

$$\sigma_n(l) = \frac{\cos(\lambda_n l) - \cosh(\lambda_n l)}{\sin(\lambda_n x) - \sinh(\lambda_n x)}$$

To homogeneous simply supported beam, its eigenvalue and eigenvector are

$$\omega_n = \frac{n^2\pi^2}{L^2}\sqrt{\frac{EI}{m}}, \quad n = 1, 2, \ldots, \infty \tag{6.36}$$

$$\varphi_n(x) = \sin\left(\frac{n\pi x}{L}\right) \tag{6.37}$$

When the eigenvectors of a matrix correspond to different eigenvalue, the eigenvectors must be orthogonality. The following will discuss the conditions for orthogonality. We know

$$EI\ddot{\varphi}_n = m\omega_n^2\varphi_n$$
$$EI\ddot{\varphi}_k = m\omega_k^2\varphi_k$$
$$\int_0^L \varphi_k EI\ddot{\varphi}_n \mathrm{d}x = \int_0^L m\omega_n^2\varphi_k\varphi_n \mathrm{d}x$$
$$\int_0^L \varphi_n EI\ddot{\varphi}_k \mathrm{d}x = \int_0^L m\omega_k^2\varphi_k\varphi_n \mathrm{d}x$$

and

$$\int_0^L \varphi_n EI\ddot{\varphi}_k \mathrm{d}x = \varphi_k[EI\ddot{\varphi}_k]_0^L - \int_0^L \dot{\varphi}_n(EI\ddot{\varphi}_k)\mathrm{d}x$$

$$= \{\varphi_k[EI\ddot{\varphi}_k]\}_0^L - \{\dot{\varphi}_n[EI\ddot{\varphi}_k]\}_0^L + \int_0^L \ddot{\varphi}_n EI\ddot{\varphi}_k \mathrm{d}x$$

We deduce

$$\begin{cases} \int_0^L EI(x)\ddot{\varphi}_n(x)\ddot{\varphi}_k(x)\mathrm{d}x = 0, & \text{for } n \neq k \\ \int_0^L m(x)\varphi_n(x)\varphi_k(x)\mathrm{d}x = 0, & \text{for } n \neq k \end{cases} \tag{6.38}$$

4. **Response Analysis Using Eigenfunction Expansion Under Force Excitation**

When a continuous beam is subjected to a force excitation, Eq. (6.38) becomes Eq. (6.39), with the boundary conditions remaining the same as in Eq. (6.31).

$$\frac{\partial^2}{\partial x^2}\left[EI(x)\frac{\partial^2 y}{\partial x^2} + \varepsilon(x)\frac{\partial^3 y}{\partial x^2 \partial t}\right] + m(x)\ddot{y} + c(x)\dot{y} = f(x, t) \tag{6.39}$$

Initial conditions are

$$y_0(x) = y(x, 0), \quad \dot{y}_0(x) = \dot{y}(x, 0)$$

The boundary conditions are the same as Eq. (6.31)'s. The response $y(x, t)$ is

$$y(x, t) = \sum_{n=1}^{\infty} a_n(t)\varphi_n(x)$$

Its eigenfunction is

$$E\ddot{I}\ddot{\varphi}_n = m\omega_n^2\varphi_n(x)$$

under

$$\int_0^L EI\ddot{\varphi}_n\ddot{\varphi}_k\mathrm{d}x = 0, \quad \int_0^L m\varphi_n\varphi_k\mathrm{d}x = 0, \quad n \neq k$$

Equation (6.39) becomes

$$\left[EI\sum_{n=1}^{\infty} a_n(t)\ddot{\varphi}_n + \varepsilon\sum_{n=1}^{\infty}\dot{a}_n(t)\ddot{\varphi}_n\right]'' + c(x)\sum_{n=1}^{\infty}\dot{a}_n(t)\varphi_n + m(x)\sum_{n=1}^{\infty}\ddot{a}_n(t)\varphi_n = f(x, t)$$

$$\sum_{n=1}^{\infty} a_n(t)(EI\ddot{\varphi}_n) + \sum_{n=1}^{\infty}\dot{a}_n(t)[\varepsilon(\ddot{x})\ddot{\varphi}_n] + c(x)\sum_{n=1}^{\infty}\dot{a}_n(t)\varphi_n + m(x)\sum_{n=1}^{\infty}\ddot{a}_n(t)\varphi_n = f(x, t)$$

At last, we get

$$\sum_{n=1}^{\infty} a_n(t) m\omega_n^2 \varphi_n + \sum_{n=1}^{\infty} \dot{a}_n(t)[\varepsilon(\ddot{x})\ddot{\varphi}_n] + c(x) \sum_{n=1}^{\infty} \dot{a}_n(t)\varphi_n + m(x) \sum_{n=1}^{\infty} \ddot{a}_n(t)\varphi_n = f(x,t)$$

In Eq. (6.39), $\varepsilon(x)$ is proportional damping model.
When the damping force proportional to time rate of bending strain, $\varepsilon(x)$ is

$$\varepsilon(x) = \nu EI(x)$$

When the damping force proportional to velocity, $\varepsilon(x)$ is

$$\varepsilon(x) = \alpha m(x)$$

Equation (6.39) becomes

$$\sum_{n=1}^{\infty} a_n(t) m\omega_n^2 \varphi_n + \nu \sum_{n=1}^{\infty} \dot{a}_n(t)[EI(\ddot{x})\ddot{\varphi}_n] + \alpha m(x) \sum_{n=1}^{\infty} \dot{a}_n(t)\varphi_n + m(x) \sum_{n=1}^{\infty} \ddot{a}_n(t)\varphi_n = f(x,t)$$

$$\sum_{n=1}^{\infty} a_n(t) m\omega_n^2 \varphi_n + \nu \sum_{n=1}^{\infty} \dot{a}_n(t) m\omega_n^2 \varphi_n(x) + \alpha m(x) \sum_{n=1}^{\infty} \dot{a}_n(t)\varphi_n + m(x) \sum_{n=1}^{\infty} \ddot{a}_n(t)\varphi_n = f(x,t)$$

$$\sum_{n=1}^{\infty} a_n(t)\omega_n^2 \int_0^L m(x)\varphi_k\varphi_n \mathrm{d}x + \nu \sum_{n=1}^{\infty} \dot{a}_n(t)\omega_n^2 \int_0^L m(x)\varphi_k\varphi_n \mathrm{d}x + \sum_{n=1}^{\infty} \dot{a}_n(t) \int_0^L c(x)\varphi_k\varphi_n \mathrm{d}x$$
$$+ \sum_{n=1}^{\infty} \ddot{a}_n(t) \int_0^L m(x)\varphi_k\varphi_n \boldsymbol{d}x = \int_0^L \varphi_k(x) f(x,t)\mathrm{d}x$$

We deduce

$$m_n\ddot{a}_n + m_n(\alpha + \nu\omega_n^2)\dot{a}_n + m_n\omega_n^2 a_n = \overline{p}_n(t)$$
$$\ddot{a}_n + (\alpha + \nu\omega_n^2)\dot{a}_n + \omega_n^2 a_n = \frac{\overline{p}_n(t)}{m_n} = p_n(t)$$
$$\ddot{a}_n + 2\eta_n\omega_n\dot{a}_n + \omega_n^2 a_n = p_n(t)$$
$$2\eta_n\omega_n = \alpha + \nu\omega_n^2$$

here

$$p_n(t) = \frac{\int_0^L \varphi_n(x) f(x,t)\mathrm{d}x}{\int_0^L \varphi_n^2(x) m(x)\mathrm{d}x}, \quad n = 1, 2, \ldots, \infty$$

From

$$y(x,t) = \sum_{k=1}^{\infty} a_k(t)\varphi_k(x)$$

Initial conditions are

$$y(x,0) = \sum_{k=1}^{\infty} a_k(0)\varphi_k(x)$$

$$a_n(0) = \int_0^L m(x)\varphi_n(x)y(x,0)\mathrm{d}x$$

$$\dot{a}_n(0) = \int_0^L m(x)\varphi_n(x)\dot{y}(x,0)\mathrm{d}x$$

The final solution $y(x, t)$ is

$$y(x,t) = \sum_{k=1}^{\infty} \varphi_n(x)\left\{ \mathrm{e}^{-\eta_n\omega_n t}[A_n\cos(\omega_{dn}t) + B_n\sin(\omega_{dn}t)] + \int_0^t h_n(t-\tau)p_n(\tau)\mathrm{d}\tau \right\}$$

Once the displacement field is found, we can easily find stress resultants and the required bending/shear stresses.

Example 6.2 An undamped simply supported beam carries a united distribution load q. The load is suddenly removed. Determine the ensuing vibrations.

Solution:

The initial deflection of the beam satisfies the equation

$$EIy_0^{iv} = q$$

Assume that the initial velocity of the beam is zero.

$$y(x,t) = \sum_{n=1}^{\infty} a_n(t)\varphi_n(x)$$

$$\ddot{a}_n + \omega_n^2 a_n = 0$$

$$\Rightarrow a_n(t) = A_n\cos(\omega_n t) + B_n\sin(\omega_n t)$$

$$\dot{y}(x,0)=0\Rightarrow \dot{a}_n(0)=0\Rightarrow y(x,t)=\sum_{n=1}^{\infty}A_n\cos(\omega_n t)\sin\left(\frac{n\pi x}{L}\right)$$

$$y(x,0)=\sum_{n=1}^{\infty}A_n\sin\left(\frac{n\pi x}{L}\right)=y_0(x)$$

$$EIy_0^{iv}=q\Rightarrow\sum_{n=1}^{\infty}EI\left(\frac{n\pi}{L}\right)^4A_n\sin\left(\frac{n\pi x}{L}\right)=q\Rightarrow A_n=\frac{L^4}{n^5\pi^5}\left(\frac{q}{EI}\right)$$

$$y(x,t)=\sum_{n=1,3,5}^{\infty}\frac{4L^2}{n^5\pi^5}\left(\frac{q}{EI}\right)\cos(\omega_n t)\sin\left(\frac{n\pi x}{L}\right)$$

Example 6.3 An united distribution load $qU(t)$ is suddenly applied on an undamped simply supported beam. Determine the ensuing vibrations, assume that the beam is at rest at $t=0$.

Solution:

$$EIy_0^{iv}+m\ddot{y}=qU(t)$$

$$\ddot{a}_n+\omega_n^2a_n=\frac{q}{m_n}\int_0^L U(t)q\sin\left(\frac{n\pi x}{L}\right)\mathrm{d}x$$

$$y(x,t)=\sum_{n=1,3,5}^{\infty}\frac{2qL}{n\pi m_n\omega_n^2}[1-\cos(\omega_n t)]\sin\left(\frac{n\pi x}{L}\right)$$

6.7 MDOF System Response Under Random Excitation

1. **Under Stationary Vector Random Excitation Response Analysis**

Random force $\boldsymbol{F}(t)$ excites a MDOF system, and its kinematic equation is

$$\boldsymbol{M}\ddot{\boldsymbol{X}}+\boldsymbol{C}\dot{\boldsymbol{X}}+\boldsymbol{K}\boldsymbol{X}=\boldsymbol{F}(t) \tag{6.40}$$

Initial conditions are

$$\boldsymbol{X}(0) = \boldsymbol{X}_0, \quad \dot{\boldsymbol{X}}(0) = \dot{\boldsymbol{X}}_0$$

Here, $\boldsymbol{X}(t)$ is a $N \times 1$ vector. $\boldsymbol{F}(t)$ satisfies

$$\langle \boldsymbol{F}(t) \rangle = 0, \quad \langle \boldsymbol{F}(t)\boldsymbol{F}^{\mathrm{T}}(t+\tau) \rangle = \boldsymbol{R}_{FF}(\tau)$$

$\boldsymbol{R}_{FF}(\tau)$ is a $N \times N$ matrix.

The response of the system in the steady state is

$$\boldsymbol{X}_T(\omega) = \boldsymbol{H}(\omega)\boldsymbol{F}_T(\omega) \tag{6.41}$$

The PSD of the response is

$$\begin{aligned} \boldsymbol{S}_{XX}(\omega) &= \lim_{T\to\infty} \frac{1}{T}\left\langle \boldsymbol{X}_T(\omega)\boldsymbol{X}_T^{*\mathrm{T}}(\omega) \right\rangle \\ &= \lim_{T\to\infty} \frac{1}{T}\left\langle \boldsymbol{H}(\omega)\boldsymbol{F}_T(\omega)\boldsymbol{F}_T^{*\mathrm{T}}(\omega)\boldsymbol{H}^{*\mathrm{T}}(\omega) \right\rangle \\ &= \boldsymbol{H}(\omega) \lim_{T\to\infty} \frac{1}{T}\left\langle \boldsymbol{F}_T(\omega)\boldsymbol{F}_T^{*\mathrm{T}}(\omega) \right\rangle \boldsymbol{H}^{*\mathrm{T}}(\omega) \\ &= \boldsymbol{H}(\omega)\boldsymbol{S}_{FF}(\omega)\boldsymbol{H}^{*\mathrm{T}}(\omega) \end{aligned} \tag{6.42}$$

When $\boldsymbol{X}(0) = \mathbf{0}, \quad \dot{\boldsymbol{X}}(0) = \mathbf{0}$, get

$$\boldsymbol{X}(t) = \int_0^t [\boldsymbol{h}(t-\tau)]\boldsymbol{F}(\tau)\boldsymbol{d}\tau \tag{6.43}$$

$$\langle \boldsymbol{X}(t) \rangle = \int_0^t [\boldsymbol{h}(t-\tau)]\langle \boldsymbol{F}(\tau) \rangle \boldsymbol{d}\tau = 0 \tag{6.44}$$

Set

$$\boldsymbol{X}(t_1) = \int_0^{t_1} [\boldsymbol{h}(t_1-\tau_1)]\boldsymbol{F}(\tau_1)\mathrm{d}\tau_1$$

$$\boldsymbol{X}(t_2) = \int_0^{t_2} [\boldsymbol{h}(t_2-\tau_2)]\boldsymbol{F}(\tau_2)\mathrm{d}\tau_2$$

therefore

$$\boldsymbol{X}^{\mathrm{T}}(t_2) = \int_0^{t_2} \boldsymbol{F}^{\mathrm{T}}(\tau_2)[\boldsymbol{h}(t_2 - \tau_2)]^{\mathrm{T}} \mathrm{d}\tau_2$$

We get

$$\langle X(t_1)X^T(t_2)\rangle = \int_0^{t_1}\int_0^{t_2} [h(t_1 - \tau_1)]\langle F(\tau_1)F^t(\tau_2)\rangle[h(t_2 - \tau_2)]^T d\tau_1 d\tau_2$$

$$R(t_1, t_2) = \int_0^{t_1}\int_0^{t_2} [h(t_1 - \tau_1)]R_{FF}(\tau_1, \tau_2)[h(t_2 - \tau_2)]^T d\tau_1 d\tau_2$$

$$= \int_0^{t_1}\int_0^{t_2} [h(t_1 - \tau_1)]R_{FF}(\tau_2 - \tau_1)[h(t_2 - \tau_2)]^T d\tau_1 d\tau_2$$

Suppose the system starts from rest, and $\boldsymbol{X}(t) = \boldsymbol{\Phi Z}(t)$. We deduce

$$X_k(t) = \sum_{n=1}^{N} \Phi_{kn} Z_n(t) \tag{6.45}$$

and get

$$\ddot{Z}_n + 2\eta_n\omega_n\dot{Z}_n + \omega_n^2 Z_n = p_n(t) \tag{6.46}$$

$$p_n(t) = \sum_{s=1}^{N} \Phi_{ns}^{\mathrm{T}} F_s(t) = \sum_{s=1}^{N} \Phi_{sn} F_s(t) \tag{6.47}$$

From Eqs. (6.46) and (6.47), find

$$Z_n(t) = \int_0^t h_n(t - \tau)\left[\sum_{s=1}^{N} \Phi_{sn} F_s(\tau)\right] \mathrm{d}\tau \tag{6.48}$$

$$X_n(t) = \sum_{n=1}^{N}\sum_{s=1}^{N} \Phi_{kn}\Phi_{sn} \int_0^t h_n(t - \tau)\left[\sum_{s=1}^{N} F_s(\tau)\right] \mathrm{d}\tau \tag{6.49}$$

Based on Eq. (6.49), mean covariance and other moments of $\boldsymbol{X}(t)$ can be evaluated.

2. **MDOF Systems Under Random Support Motions**

Taking the building frame in Fig. 6.4 as an example, we discuss the response of a MDOF system under support motion. If the support motion is uniform, its equivalent dynamic model is given in Fig. 6.4.

(1) Structures under same support motions.

The expectation and autocorrelation of the excitation $u(t)$ are

$$\langle u(t) \rangle = 0, \quad \langle u(t)u(t+\tau) \rangle = R_{uu}(\tau)$$

The kinematic equations of Fig. 6.4 can be deduced as the same as the characteristic of the structure in Fig. 6.4.

Compare with Eq. (6.40), we can obtain

$$\begin{aligned} \boldsymbol{F}(t) &= -\boldsymbol{MI}\ddot{\boldsymbol{u}}(t) \\ \boldsymbol{F}_T(\omega) &= -\boldsymbol{MI}\ddot{\boldsymbol{U}}(\omega) \end{aligned}$$

therefore

$$\begin{aligned} \boldsymbol{S}_{ff}(\omega) &= \boldsymbol{MI}\langle \ddot{\boldsymbol{U}}(\omega)\ddot{\boldsymbol{U}}_T^*(\omega) \rangle \boldsymbol{I}^{\mathrm{T}}\boldsymbol{M} = \boldsymbol{MIS}_{uu}(\omega)\boldsymbol{I}^{\mathrm{T}}\boldsymbol{M} \\ \boldsymbol{S}_{xx}(\omega) &= \boldsymbol{H}(\omega)\boldsymbol{S}_{ff}(\omega)\boldsymbol{H}^{*\mathrm{T}}(\omega) \\ \boldsymbol{S}_{xx}(\omega) &= \boldsymbol{H}(\omega)\boldsymbol{MIS}_{uu}(\omega)\boldsymbol{I}^{\mathrm{T}}\boldsymbol{MH}^{*\mathrm{T}}(\omega) \end{aligned}$$

and

$$\boldsymbol{H}(\omega) = (-\omega^2\boldsymbol{M} + \mathrm{i}\omega\boldsymbol{C} + \boldsymbol{K})^{-1}$$

When $\langle \boldsymbol{F}(t) \rangle = -\boldsymbol{MI}\langle \ddot{\boldsymbol{u}}(t) \rangle = 0$, we know

$$\begin{aligned} \boldsymbol{R}_{ff}(t_1, t_2) &= \left\langle \boldsymbol{F}(t_1)\boldsymbol{F}^{\mathrm{T}}(t_2) \right\rangle (N \times N) \\ &= \boldsymbol{MI}\langle \ddot{\boldsymbol{u}}(t_1)\ddot{\boldsymbol{u}}(t_2) \rangle \boldsymbol{I}^{\mathrm{T}}\boldsymbol{M} \\ &= \boldsymbol{MIR}_{\ddot{u}\ddot{u}}(t_2, t_1)\boldsymbol{I}^{\mathrm{T}}\boldsymbol{M} \end{aligned}$$

The response $\boldsymbol{X}(t)$ can be deduced

$$\boldsymbol{X}(t) = \int_0^t [\boldsymbol{h}(t-\tau)]\boldsymbol{F}(\tau)\boldsymbol{d}\tau = -\int_0^t [\boldsymbol{h}(t-\tau)]\boldsymbol{MI}\dot{\boldsymbol{u}}(\tau)\boldsymbol{d}\tau$$

The expectation of $\boldsymbol{X}(t)$ is

$$E[\boldsymbol{X}(t)] = -\int_0^t [\boldsymbol{h}(t-\tau)]\boldsymbol{MI}\langle\dot{\boldsymbol{u}}(\tau)\rangle \boldsymbol{d}\tau = 0$$

And the autocorrelation function of $\boldsymbol{X}(t)$ is

$$\boldsymbol{R}_{xx}(t_1, t_2) = -\int_0^{t_1}\int_0^{t_2} [\boldsymbol{h}(t_1-\tau_1)]\boldsymbol{MIR}_{uu}(\tau_1, \tau_2)\boldsymbol{I}^{\mathrm{T}}\boldsymbol{M}[\boldsymbol{h}(t_1-\tau_1)]^t \mathrm{d}\tau_1\mathrm{d}\tau_2$$

From $x(t) = \boldsymbol{\Phi y}(t)$, we can get

$$x_k(t) = \sum_{r=1}^{n} \Phi_{kr} y_r(t) \tag{6.50}$$

hence

$$\ddot{y}_n + 2\xi_n\omega_n\dot{y}_n + \omega_n^2 y_n = p_n(t) \tag{6.51}$$

The modal participation factor

$$\{\boldsymbol{\gamma}\} = -\boldsymbol{\Phi}^{\mathrm{T}}\boldsymbol{MI}$$

Equation (6.51) becomes

$$\ddot{y}_n + 2\xi_n\omega_n\dot{y}_n + \omega_n^2 y_n = \gamma_n \ddot{u}(t) \tag{6.52}$$

The solution $y_n(t)$ of Eq. (6.52) is

$$y_n(t) = \int_0^t h_n(t-\tau)\gamma_n\dot{u}(\tau)\mathrm{d}\tau \tag{6.53}$$

From Eq. (6.50) and Eq. (6.53), we get

$$x_k(t) = \sum_{n=1}^{N} \Phi_{kn} \int_0^t h_n(t-\tau)\gamma_n\dot{u}(\tau)\mathrm{d}\tau$$

$$\langle x_k(t)\rangle = \sum_{n=1}^{N} \Phi_{kn} \int_0^t h_n(t-\tau)\gamma_n\langle\dot{u}(\tau)\rangle\mathrm{d}\tau$$

and

$$\begin{aligned}\langle x_k(t_1)x_k(t_2)\rangle &= \sum_{n=1}^{N}\sum_{m=1}^{N}\Phi_{kn}\Phi_{km}\langle y_n(t_1)y_m(t_2)\rangle \\ &= \sum_{n=1}^{N}\sum_{m=1}^{N}\Phi_{kn}\Phi_{km}\int_0^{t_1}\int_0^{t_2} h_n(t_1-\tau_1)h_n(t_2-\tau_2)\gamma_n\gamma_m\langle\ddot{u}(\tau_1)\ddot{u}(\tau_2)\rangle \mathrm{d}\tau_1\mathrm{d}\tau_2 \\ &= \sum_{n=1}^{N}\sum_{m=1}^{N}\Phi_{kn}\Phi_{km}\int_0^{t_1}\int_0^{t_2} h_n(t_1-\tau_1)h_n(t_2-\tau_2)\gamma_n\gamma_m R_{uu}(\tau_1,\tau_2)\mathrm{d}\tau_1\mathrm{d}\tau_2 \\ &= \sum_{n=1}^{N}\sum_{m=1}^{N}\Phi_{kn}\Phi_{km}\int_0^{t_1}\int_0^{t_2} h_n(t_1-\tau_1)h_n(t_2-\tau_2)\gamma_n\gamma_m R_{uu}(\tau_1-\tau_2)\mathrm{d}\tau_1\mathrm{d}\tau_2\end{aligned}$$

The variance is

$$\sigma_x^2(t) = \sum_{n=1}^{N}\sum_{m=1}^{N}\Phi_{kn}\Phi_{km}\int_0^{t_1}\int_0^{t_2} h_n(t_1-\tau_1)h_n(t_2-\tau_2)\gamma_n\gamma_m R_{uu}(\tau_1-\tau_2)\mathrm{d}\tau_1\mathrm{d}\tau_2$$

From $x(t) = \boldsymbol{\Phi}\mathbf{y}(t), \quad x_k(t) = \sum_{r=1}^{n}\Phi_{kr}y_r(t)$, we get

$$X_{kT}(\omega) = \sum_{n=1}^{N}\Phi_{kn}Y_{nT}(\omega)$$

From $\ddot{y}_n + 2\xi_n\omega_n\dot{y}_n + \omega_n^2 y_n = \gamma_n\ddot{u}(t), \quad Y_{nT}(\omega) = \frac{\gamma_n\ddot{U}_T(\omega)}{\omega_n^2-\omega^2+\mathrm{i}2\xi_n\omega_n\omega}$, we get

$$X_{nT}(\omega) = \sum_{n=1}^{N}\Phi_{kn}\frac{\gamma_n\ddot{U}_T(\omega)}{\omega_n^2-\omega^2+\mathrm{i}2\xi_n\omega_n\omega}$$

The PSD of $\boldsymbol{X}(t)$ is

$$\begin{aligned}S_{x_kx_k}(\omega) &= \sum_{n=1}^{N}\sum_{m=1}^{N}\frac{\Phi_{kn}\Phi_{km}\gamma_n\gamma_m}{\left(\omega_n^2-\omega^2+\mathrm{i}2\eta_n\omega_n\omega\right)\left(\omega_m^2-\omega^2+\mathrm{i}2\eta_m\omega_m\omega\right)}S_{UU}(\omega) \\ &= \sum_{n=1}^{N}\frac{\Phi_{kn}^2\gamma_n^2 S_{UU}(\omega)}{\left(\omega_n^2-\omega^2\right)^2+(2\eta_n\omega_n\omega)^2} \\ &+ \sum_{n=1}^{N}\sum_{\substack{m=1\\ n\neq m}}^{N}\frac{\Phi_{kn}\Phi_{km}\gamma_n\gamma_m}{\left(\omega_n^2-\omega^2+\mathrm{i}2\eta_n\omega_n\omega\right)\left(\omega_m^2-\omega^2+\mathrm{i}2\eta_m\omega_m\omega\right)}S_{UU}(\omega)\end{aligned}$$

The variance of $X(t)$ is

$$\sigma_{x_k}^2(t) = \int_0^\infty S_{x_k x_k}(\omega)\mathrm{d}\omega$$

The contribution of modal interactions and correction due to modal interaction is ignored at the above analysis.

(2) Structures under differential support motions

When the system is applied support motions and support motions is differential, we can build the kinematic matrix as follows

$$\begin{pmatrix} \boldsymbol{M} & \boldsymbol{M}_g \\ \boldsymbol{M}_g^T & \boldsymbol{M}_{gg} \end{pmatrix}\begin{pmatrix} \ddot{\boldsymbol{u}}^T \\ \ddot{\boldsymbol{u}}_g \end{pmatrix} + \begin{pmatrix} \boldsymbol{C} & \boldsymbol{C}_g \\ \boldsymbol{C}_g^T & \boldsymbol{C}_{gg} \end{pmatrix}\begin{pmatrix} \dot{\boldsymbol{u}}^T \\ \dot{\boldsymbol{u}}_g \end{pmatrix} + \begin{pmatrix} \boldsymbol{K} & \boldsymbol{K}_g \\ \boldsymbol{K}_g^T & \boldsymbol{K}_{gg} \end{pmatrix}\begin{pmatrix} \boldsymbol{u}^T \\ \boldsymbol{u}_g \end{pmatrix} = \begin{pmatrix} \boldsymbol{0} \\ \boldsymbol{p}_g(t) \end{pmatrix} \tag{6.54}$$

We can divide Eq. (6.54) into two parts, pseudo-dynamic response and dynamic response.

We know that

$$\boldsymbol{H}(\omega) = \left(-\omega^2\boldsymbol{M} + \mathrm{i}\omega\boldsymbol{C} + \boldsymbol{K}\right)^{-1} = \sum_{n=1}^{N} \frac{\Phi_{rn}\Phi_{sn}}{\omega_n^2 - \omega^2 + \mathrm{i}2\eta_n\omega_n\omega}$$

$$\boldsymbol{h}(t) = \boldsymbol{h}_{rs}(t) = \sum_{n=1}^{N} \Phi_{rn}\Phi_{sn}\frac{1}{\omega_{dn}}\exp(-\eta_n\omega_n t)\sin(\omega_{dn}t)$$

$$H_{ij}(\omega) = \int_{-\infty}^{\infty} h_{ij}(t)\exp(\mathrm{i}\omega t)\mathrm{d}t$$

$$h_{ij}(t) = \frac{1}{2\pi}\int_{-\infty}^{\infty} H_{ij}(\omega)\exp(-\mathrm{i}\omega t)\mathrm{d}\omega$$

$$S_{xx}(\omega) = \boldsymbol{H}(\omega)\boldsymbol{S}_{ff}(\omega)\boldsymbol{H}^{*\mathrm{T}}(\omega)$$

$$R_{xx}(t_1, t_2) = \int_0^{t_1}\int_0^{t_2} [h(t_1 - \tau_1)]R_{FF}(\tau_2 - \tau_1)[h(t_2 - \tau_2)]^{\mathrm{T}}\mathrm{d}\tau_1\mathrm{d}\tau_2$$

The pseudo-dynamic response is

$$\begin{pmatrix} \boldsymbol{K} & \boldsymbol{K}_g \\ \boldsymbol{K}_g^{\mathrm{T}} & \boldsymbol{K}_{gg} \end{pmatrix}\begin{pmatrix} \boldsymbol{u}^p \\ \boldsymbol{u}_g \end{pmatrix} = \begin{pmatrix} \boldsymbol{0} \\ \boldsymbol{p}_g^p(t) \end{pmatrix}$$
$$\boldsymbol{K}\boldsymbol{u}^p + \boldsymbol{K}_g\boldsymbol{u}_g = 0 \Rightarrow u^p = -\boldsymbol{K}^{-1}\boldsymbol{K}_g\boldsymbol{u}_g = -\boldsymbol{\Gamma}\boldsymbol{u}_g$$
$$\boldsymbol{\Gamma} = \boldsymbol{K}^{-1}\boldsymbol{K}_g$$
$$\boldsymbol{K}_g^T\boldsymbol{u}^p + \boldsymbol{K}_{gg}\boldsymbol{u}_g = \boldsymbol{p}_g^p(t)$$

The total response includes the pseudo-dynamic response and the dynamic response,

$$\begin{aligned} \begin{pmatrix} \boldsymbol{u}^{\mathrm{T}} \\ \boldsymbol{u}_g \end{pmatrix} &= \begin{pmatrix} \boldsymbol{u}^p(t) \\ \boldsymbol{u}_g(t) \end{pmatrix} + \begin{pmatrix} \boldsymbol{u}(t) \\ \boldsymbol{0} \end{pmatrix}\begin{pmatrix} \boldsymbol{M} & \boldsymbol{M}_g \\ \boldsymbol{M}_g^{\mathrm{T}} & \boldsymbol{M}_{gg} \end{pmatrix}\begin{pmatrix} \ddot{\boldsymbol{u}} + \ddot{\boldsymbol{u}}^p(t) \\ \ddot{\boldsymbol{u}} \end{pmatrix} \\ &+ \begin{pmatrix} \boldsymbol{C} & \boldsymbol{C}_g \\ \boldsymbol{C}_g^{\mathrm{T}} & \boldsymbol{C}_{gk} \end{pmatrix}\begin{pmatrix} \dot{\boldsymbol{u}} + \dot{\boldsymbol{u}}^p(t) \\ \dot{\boldsymbol{u}}_g \end{pmatrix} + \begin{pmatrix} \boldsymbol{K} & \boldsymbol{K}_g \\ \boldsymbol{K}_g^{\mathrm{T}} & \boldsymbol{K}_{gg} \end{pmatrix}\begin{pmatrix} \boldsymbol{u} + \boldsymbol{u}^p \\ \boldsymbol{u}_g \end{pmatrix} \end{aligned} \tag{6.55}$$

We can deduce

$$\boldsymbol{M}\ddot{\boldsymbol{u}} + \boldsymbol{C}\dot{\boldsymbol{u}} + \boldsymbol{K}\boldsymbol{u} = \boldsymbol{p}_{\mathrm{eff}}(t) \tag{6.56}$$

where

$$\begin{aligned} \boldsymbol{p}_{\mathrm{eff}}(t) &= -\boldsymbol{M}\ddot{\boldsymbol{u}}\boldsymbol{p}(t) - \boldsymbol{M}_g\ddot{\boldsymbol{u}}_g - \boldsymbol{C}\dot{\boldsymbol{u}}\boldsymbol{p}(t) - \boldsymbol{C}_g\dot{\boldsymbol{u}}_g \\ &= -\boldsymbol{M}\boldsymbol{\Gamma}\ddot{\boldsymbol{u}}_g(t) - \boldsymbol{M}_g\ddot{\boldsymbol{u}}_g - \boldsymbol{C}\boldsymbol{\Gamma}\dot{\boldsymbol{u}}_g(t) - \boldsymbol{C}_g\dot{\boldsymbol{u}}_g \\ &= -(\boldsymbol{M}\boldsymbol{\Gamma} + \boldsymbol{M}_g)\ddot{\boldsymbol{u}}_g(t) - (\boldsymbol{C}\boldsymbol{\Gamma} + \boldsymbol{C}_g)\dot{\boldsymbol{u}}_g(t) \end{aligned}$$

At last, Eq. (6.56) becomes

$$\boldsymbol{M}\ddot{\boldsymbol{u}} + \boldsymbol{C}\dot{\boldsymbol{u}} + \boldsymbol{K}\boldsymbol{u} = -(\boldsymbol{M}\boldsymbol{\Gamma} + \boldsymbol{M}_g)\ddot{\boldsymbol{u}}_g(t) - (\boldsymbol{C}\boldsymbol{\Gamma} + \boldsymbol{C}_g)\dot{\boldsymbol{u}}_g(t) \tag{6.57}$$

There is a special case, mass matrix $\boldsymbol{M}_g$ is diagonal, get $\boldsymbol{M}_g = 0$; $\boldsymbol{C}$ is proportional to K, get $\boldsymbol{C} = \alpha\boldsymbol{K}$.

Therefore

$$\boldsymbol{C}\boldsymbol{\Gamma} + \boldsymbol{C}_g = \alpha(\boldsymbol{K}\boldsymbol{\Gamma} + \boldsymbol{K}_g) = \alpha(\boldsymbol{K}\boldsymbol{K}^{-1}\boldsymbol{K}_g + \boldsymbol{K}_g) = 0$$
$$\boldsymbol{M}\ddot{\boldsymbol{u}} + \boldsymbol{C}\dot{\boldsymbol{u}} + \boldsymbol{K}\boldsymbol{u} = -\boldsymbol{M}\boldsymbol{\Gamma}\ddot{\boldsymbol{u}}_g(t)$$

If all supports suffer the same motion, $N_g = 1$(N_g refers to the number of DOF of the support (ground) motion or the number of support points), $\boldsymbol{\Gamma} = (\,1\;1\;\cdots\;1\,)^{\mathrm{T}}$.

3. **Random Vibration Analysis in Time Domain**

When a MDOF system is subjected to support motions, the kinematic equation can be deduced as Eq. (6.58), with the auto-covariance matrix of the stationary random process.

$$\boldsymbol{M}\ddot{\boldsymbol{u}} + \boldsymbol{C}\dot{\boldsymbol{u}} + \boldsymbol{K}\boldsymbol{u} = -(\boldsymbol{M}\boldsymbol{\Gamma} + \boldsymbol{M}_g)\ddot{\boldsymbol{u}}_g(t) - (\boldsymbol{C}\boldsymbol{\Gamma} + \boldsymbol{C}_g)\dot{\boldsymbol{u}}_g(t) = \boldsymbol{p}(t) \tag{6.58}$$

$\boldsymbol{u}_g(t)$ is a $N_g \times 1$ vector of stationary random process with zero mean and auto-covariance matrix.

$$\boldsymbol{R}_{gg}(t_1, t_2) = \left\langle \boldsymbol{u}_g(t_1)\boldsymbol{u}_g^{\mathrm{T}}(t_2)\right\rangle \boldsymbol{R}_{gg}(t_1 - t_2) \tag{6.59}$$

Therefore, we can get

$$\begin{aligned}
\boldsymbol{p}(t_1) &= -(\boldsymbol{M}\boldsymbol{\Gamma} + \boldsymbol{M}_g)\ddot{\boldsymbol{u}}_g(t_1) - (\boldsymbol{C}\boldsymbol{\Gamma} + \boldsymbol{C}_g)\dot{\boldsymbol{u}}_g(t_1) \\
\boldsymbol{p}^{\mathrm{T}}(t_2) &= -\ddot{\boldsymbol{u}}_g^{\mathrm{T}}(t_2)(\boldsymbol{\Gamma}^{\mathrm{T}}\boldsymbol{M} + \boldsymbol{M}_g) - \dot{\boldsymbol{u}}_g^{\mathrm{T}}(t_2)(\boldsymbol{\Gamma}^{\mathrm{T}}\boldsymbol{C} + \boldsymbol{C}_g)
\end{aligned}$$

and

$$\begin{aligned}
&\left\langle \boldsymbol{p}(t_1)\boldsymbol{p}^{\mathrm{T}}(t_2)\right\rangle = \boldsymbol{R}_{pp}(t_1, t_2) = \left(\boldsymbol{M}\boldsymbol{\Gamma} + \boldsymbol{M}_g\right)\left\langle \ddot{\boldsymbol{u}}_g(t_1)\ddot{\boldsymbol{u}}_R^{\mathrm{T}}(t_2)\right\rangle\left(\boldsymbol{\Gamma}^{\mathrm{T}}\boldsymbol{M} + \boldsymbol{M}_g\right) \\
&+(\boldsymbol{M}\boldsymbol{\Gamma} + M_g)\left\langle \ddot{\boldsymbol{u}}_g(t_1)\ddot{\boldsymbol{u}}_g^{\mathrm{T}}(t_2)\right\rangle(\boldsymbol{\Gamma}^{\mathrm{T}}\boldsymbol{C} + \boldsymbol{M}\boldsymbol{C}_g) \\
&+(\boldsymbol{C}\boldsymbol{\Gamma} + \boldsymbol{C}_g)\left\langle \dot{\boldsymbol{u}}_g(t_1)\dot{\boldsymbol{u}}_g^{\mathrm{T}}(t_2)\right\rangle(\boldsymbol{\Gamma}^{\mathrm{T}}\boldsymbol{M} + \boldsymbol{M}_g) \\
&+(\boldsymbol{C}\boldsymbol{\Gamma} + \boldsymbol{C}_g)\left\langle \dot{\boldsymbol{u}}_g(t_1)\dot{\boldsymbol{u}}_g^{\mathrm{T}}(t_2)\right\rangle(\boldsymbol{\Gamma}^{\mathrm{T}}\boldsymbol{C} + \boldsymbol{M}\boldsymbol{C}_g)
\end{aligned}$$

The autocorrelation function of $\boldsymbol{u}_g(t)$ is

$$\boldsymbol{R}_{uu}(t_1, t_2) = \int_0^{t_1}\int_0^{t_2} [\boldsymbol{h}(t_1 - \tau_1)]\boldsymbol{R}_{pp}(\tau_2 - \tau_1)[\boldsymbol{h}(t_2 - \tau_2)]^{\mathrm{T}}\mathrm{d}\tau_1\mathrm{d}\tau_2$$

From

$$\boldsymbol{h}(t) = \boldsymbol{h}_{rs}(t) = \sum_{n=1}^{N} \Phi_{rn}\Phi_{sn}\frac{1}{\omega_{dn}}\exp(-\eta_n\omega_n t)\sin(\omega_{dn}t)$$

and

$$\begin{aligned}
\left(\left.\frac{\boldsymbol{d}^n X}{\boldsymbol{d}t^n}\right|_{t=t_1}\left.\frac{\boldsymbol{d}^m X}{\boldsymbol{d}t^m}\right|_{t=t_2}\right) &= \frac{\partial^{n+m}R_{xx}(t_1, t_2)}{\partial t_1^n \partial t_1^m} \\
\left[\frac{\mathrm{d}^n X(t+\tau)}{\mathrm{d}t^n}\frac{\mathrm{d}^m Y(t)}{\mathrm{d}t^m}\right] &= (-1)^m\frac{\mathrm{d}^{n+m}R_{xx}(\tau)}{\mathrm{d}\tau^{n+m}}
\end{aligned}$$

The total response includes the pseudo-dynamic response and the dynamic response

$$\boldsymbol{u}^{\mathrm{T}}(t) = \boldsymbol{u}^p(t) + \boldsymbol{u}(t) = \boldsymbol{\Gamma}\boldsymbol{u}_g(t) + \boldsymbol{u}(t)$$

$$= \mathbf{\Gamma} \boldsymbol{u}_g(t) + \int_0^t [\boldsymbol{h}(t-\tau)]\big[-(\boldsymbol{M\Gamma} + \boldsymbol{M}_g)\ddot{\boldsymbol{u}}_g(\tau) - (\boldsymbol{C\Gamma} + \boldsymbol{C}_g)\dot{\boldsymbol{u}}_g(\tau)\big]d\tau$$

The variance of total response includes the variance of pseudo-dynamic response and the contributions due to correlation between the pseudo-dynamic and the dynamic response.

4. **Random Vibration Analysis in Frequency Domain**

The system remains the same as in Eq. (6.59), and the PSD matrix of the stationary random process is considered.

$$\boldsymbol{S}_{gg}(\omega) = \lim_{T\to\infty} \frac{1}{T}\left\langle \boldsymbol{U}_{gT}(\omega)\boldsymbol{U}_{gT}^{*\mathrm{T}}(\omega)\right\rangle$$

we can get

$$\begin{aligned}
&\boldsymbol{p}(t) = -(\boldsymbol{M\Gamma} + \boldsymbol{M}_g)\ddot{\boldsymbol{u}}_g(t) - (\boldsymbol{C\Gamma} + \boldsymbol{C}_g)\dot{\boldsymbol{u}}_g(t) \\
&\boldsymbol{p}_T(\omega) = \omega^2(\boldsymbol{M\Gamma} + \boldsymbol{M}_g)\boldsymbol{U}_{gT}(\omega) - \mathrm{i}\omega(\boldsymbol{C\Gamma} + \boldsymbol{C}_g)\boldsymbol{U}_{gT}(\omega) \\
&= [\omega^2(\boldsymbol{M\Gamma} + \boldsymbol{M}_g) - \mathrm{i}\omega(\boldsymbol{C\Gamma} + \boldsymbol{C}_g)]\boldsymbol{U}_{gT}(\omega) \\
&\boldsymbol{p}_T^{*\mathrm{T}}(\omega) = \boldsymbol{U}_{gT}^{*\mathrm{T}}(\omega)[\omega^2(\boldsymbol{\Gamma}^{\mathrm{T}}\boldsymbol{M} + \boldsymbol{M}_g) + \mathrm{i}\omega(\boldsymbol{\Gamma}^{\mathrm{T}}\boldsymbol{C} + \boldsymbol{C}_g)]
\end{aligned}$$

$$\begin{aligned}
&\boldsymbol{S}_{pp}(\omega) = \lim_{T\to\infty} \frac{1}{T}\langle \omega^2(\boldsymbol{M\Gamma} + \boldsymbol{M}_g) - i\omega(\boldsymbol{C\Gamma} + \boldsymbol{C}_g)\rangle \boldsymbol{U}_{gT}(\omega) \\
&\boldsymbol{U}_{gT}^{*\mathrm{T}}(\omega) = \langle \omega^2(\boldsymbol{\Gamma}^{\mathrm{T}}\boldsymbol{M} + \boldsymbol{M}_g) + \mathrm{i}\omega(\boldsymbol{\Gamma}^{\mathrm{T}}\boldsymbol{C} + \boldsymbol{C}_g)\rangle \\
&\boldsymbol{S}_{pp}(\omega) = \omega^2(\boldsymbol{M\Gamma} + \boldsymbol{M}_g) - \mathrm{i}\omega(\boldsymbol{C\Gamma} + \boldsymbol{C}_g) \\
&\boldsymbol{S}_{gg}(\omega) = \omega^2(\boldsymbol{\Gamma}^{\mathrm{T}}\boldsymbol{M} + \boldsymbol{M}_g) + \mathrm{i}\omega(\boldsymbol{\Gamma}^{\mathrm{T}}\boldsymbol{C} + \boldsymbol{C}_g)
\end{aligned}$$

The PSD of $\boldsymbol{u}_R(t)$ is

$$\boldsymbol{S}_{uu}(\omega) = \boldsymbol{H}(\omega)\boldsymbol{S}_{pp}(\omega)\boldsymbol{H}^{*\mathrm{T}}(\omega)$$

Known

$$\boldsymbol{H}(\omega) = \left(-\omega^2\boldsymbol{M} + \mathrm{i}\omega\boldsymbol{C} + \boldsymbol{K}\right)^{-1} = \sum_{n=1}^{N} \frac{\boldsymbol{\Phi}_{rn}\boldsymbol{\Phi}_{sn}}{\omega_n^2 - \omega^2 + \mathrm{i}2\eta_n\omega_n\omega}$$

and the pseudo-dynamic response is

$$\begin{aligned}
&\boldsymbol{u}^p = -\boldsymbol{K}^{-1}\boldsymbol{K}_g\boldsymbol{u}_g(t) = \mathbf{\Gamma}\boldsymbol{u}_g(t) \\
&\boldsymbol{\Gamma} = -\boldsymbol{K}^{-1}\boldsymbol{K}_g \\
&\boldsymbol{S}_{pp}(\omega) = \mathbf{\Gamma}\boldsymbol{S}_{gg}(\omega)\boldsymbol{\Gamma}^{\mathrm{T}}
\end{aligned}$$

The total response

$$\begin{aligned}
&\boldsymbol{u}^{\mathrm{T}}(t) = \boldsymbol{u}^{p}(t) + \boldsymbol{u}(t) = \boldsymbol{\Gamma}\boldsymbol{u}_g(t) + \boldsymbol{u}(t)\\
&\boldsymbol{U}_T^{\mathrm{T}}(\omega) = \boldsymbol{\Gamma}\boldsymbol{U}_{gT}(\omega) + \boldsymbol{U}_T(\omega) = \boldsymbol{\Gamma}\boldsymbol{U}_{gT}(\omega) + \boldsymbol{H}(\omega)\boldsymbol{P}_T(\omega)\\
&\boldsymbol{P}_T(\omega) = \left[\omega^2(\boldsymbol{M\Gamma} + \boldsymbol{M}_g) - \mathrm{i}\omega(\boldsymbol{C\Gamma} + \boldsymbol{C}_g)\right]\boldsymbol{U}_{gT}(\omega)\\
&\boldsymbol{U}_T^{\mathrm{T}}(\omega) = \left[\boldsymbol{\Gamma} + \omega^2(\boldsymbol{M\Gamma} + \boldsymbol{M}_g) - \mathrm{i}\omega(\boldsymbol{C\Gamma} + \boldsymbol{C}_g)\right]\boldsymbol{U}_{gT}(\omega)
\end{aligned}$$

At last,

$$\begin{aligned}
&\boldsymbol{S}_{TT}(\omega) = \left[\boldsymbol{\Gamma} + \omega^2(\boldsymbol{M\Gamma} + \boldsymbol{M}_g) - \mathrm{i}\omega(\boldsymbol{C\Gamma} + \boldsymbol{C}_g)\right]\boldsymbol{S}_{gg}(\omega)\\
&\times\left[\boldsymbol{\Gamma} + \omega^2(\boldsymbol{M\Gamma} + \boldsymbol{M}_g) - \mathrm{i}\omega(\boldsymbol{C\Gamma} + \boldsymbol{C}_g)\right]^{\mathrm{T}}
\end{aligned}$$

Exercise 6

E 6.1: Figure 6.10 presents a simplified model of a dynamic vibration absorber. The main system consists of mass m_1 and spring k_1, while the secondary system includes mass m_2, spring k_2, and damper c. Assuming the interference force F acting on the main mass has a white noise spectrum with an average of 0, determine the optimal parameters of the vibration absorber that minimize the RMS value of the system response. Known $m_1, m_2, k_1, k_2, c,\ S_F(\omega) = S$, find the optimal parameters of the vibration absorber ρ_o, ξ_o.

E 6.2: In the two DOF system shown in Fig. 6.11, with the input force $F(t)$ and the output being the displacement of two masses, if the power spectral density $S_F\ (w)$ of the excitation force $F(t)$ is an ideal white noise, find the average kinetic energy of the system. Known $m, k, c, S_F(\omega) = S_0$, find the average kinetic energy of the system.

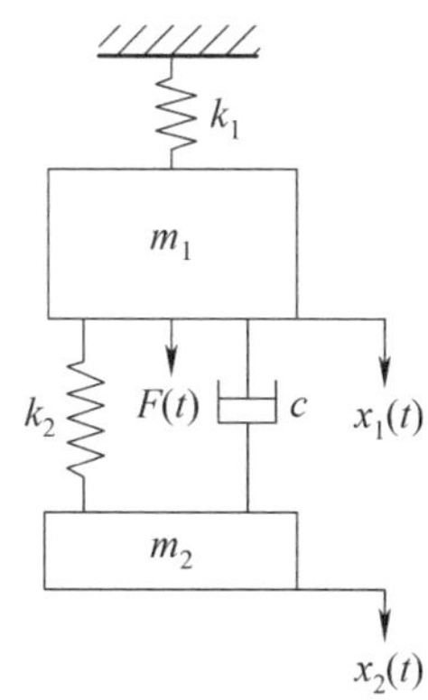

Fig. 6.10 A simplified model of a dynamic vibration absorber

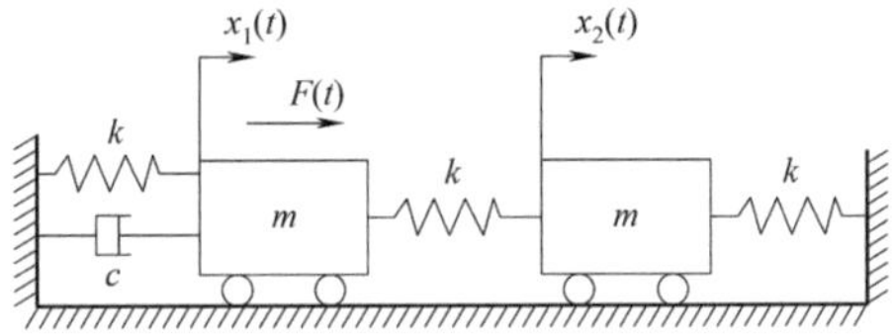

Fig. 6.11 A mechanical system with two masses

Chapter 7
Random Vibration of Railway Vehicles

In the railway vehicle dynamics, the random vibration of vehicles is a critical aspect that requires meticulous analysis and understanding. The irregularities in the track can exhibit low correlation between the important directions and the high and low spectra, as well as between the directions and the horizontal spectra. Consequently, it is feasible to disregard three types of irregular cross spectral densities in specific calculations, simplifying the analysis process.

The vertical irregularity of the track, when represented by spatial PSD, can be transformed into the PSD in the time domain, based on the operating speed v of the railway vehicle. With the knowledge of the transfer function of the vehicle under wheel excitation to the desired vehicle position, one can deduce the response PSD of the vehicle body.

For the SDOF vehicle vibration system depicted in Fig. 7.1, the track irregularity excites the vehicle, leading to forced vibration. Given the mass, damping, stiffness, and track irregularity power spectrum of the vehicle, the forced vibration equation can be formulated as follows:

$$\boldsymbol{M}\ddot{\boldsymbol{Z}} + \boldsymbol{C}\dot{\boldsymbol{Z}} + \boldsymbol{K}\boldsymbol{Z} = \boldsymbol{C}\dot{\boldsymbol{Z}}_i(t) + \boldsymbol{K}\boldsymbol{Z}_i(t) \tag{7.1}$$

Assuming the input to be exciting force, the response of the vehicle body is determined by inserting into Eq. (7.1). Here, b is complex, yielding the transfer function in the complex domain.

$$H(\omega) = \frac{b}{a} = \frac{\boldsymbol{K} + \mathrm{i}\boldsymbol{C}\omega}{-\boldsymbol{M}\omega^2 + \boldsymbol{K} + \mathrm{i}\boldsymbol{C}\omega} \tag{7.2}$$

The transfer function of the acceleration, also in the complex domain, is given by:

$$\begin{cases} H_a(\omega) = \frac{b\omega^2}{a} = \omega^2 \frac{\boldsymbol{K}+\mathrm{i}\boldsymbol{C}\omega}{-\boldsymbol{M}\omega^2+\boldsymbol{K}+\mathrm{i}\boldsymbol{C}\omega} \\ |H_a(\omega)|^2 = \omega^4 \frac{1+4\boldsymbol{D}^2\boldsymbol{r}^2}{(1-r^2)^2+4\boldsymbol{D}^2r^2} \end{cases} \tag{7.3}$$

J. Zhang and H. Zhou, *Random Vibration and Spectrum Analysis*,
https://doi.org/10.1007/978-981-95-0191-5_7

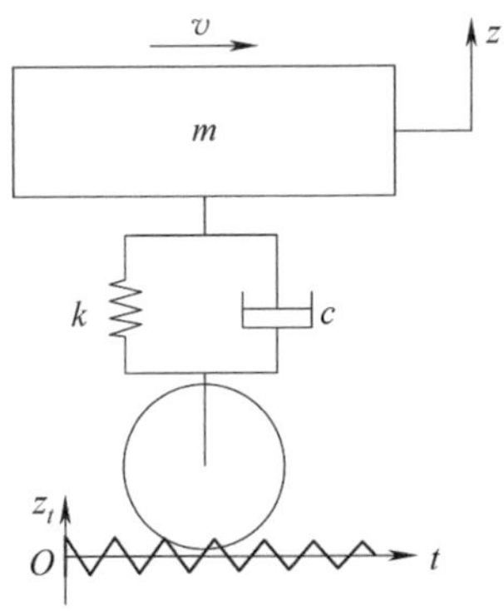

Fig. 7.1 SDOF vehicle (forced vibration)

At resonance, when the frequency of the input matches the natural frequency of the system, the absolute value of the denominator is minimized, leading to a peak in the transfer function. The acceleration power spectrum of the vehicle indicates that the response value of the railway vehicle can be reduced by diminishing the amplitude of the transfer function. This is achieved by lowering the natural frequency and employing an appropriate damping ratio D to reduce its peak value.

7.1 Vertical Random Vibration of Railway Vehicles

7.1.1 Vertical Random Response of Two-Axle Railway Vehicles

The model of a two-axle railway vehicle is illustrated in Fig. 7.2. During operation, the track vertical irregularity excitation at the front wheel is $z_{\mathrm{p}1}$, while the rear wheel excitation $z_{\mathrm{p}2}$ lagged by a certain time delay relative to the front wheel. The differential equations governing the vertical motion of the vehicle are as follows:

$$m\ddot{z} + 2c\dot{z} + 2kz = c(\dot{z}_{v1} + \dot{z}_{v2}) + k(z_{v1} + z_{v2}) \tag{7.4}$$

$$J_{\varphi}\ddot{\varphi} + 2cl^2\dot{\varphi} + 2kl^2\varphi = cl(\dot{z}_{v1} - \dot{z}_{v2}) + kl(z_{v1} - z_{v2}) \tag{7.5}$$

where, z and φ represent the float and sink vibration coordinates of the vehicle body, respectively; m and J_{φ} are the body mass and nodding moment of inertia, respectively; c and k denote the damping coefficient and spring stiffness of one end of the vehicle, respectively; $2l$ is the vehicle wheelbase, and z_{v1}, z_{v2} are the front and rear wheel vertical irregularity excitations, respectively; τ represents the delay between the rear wheel and the front wheel.

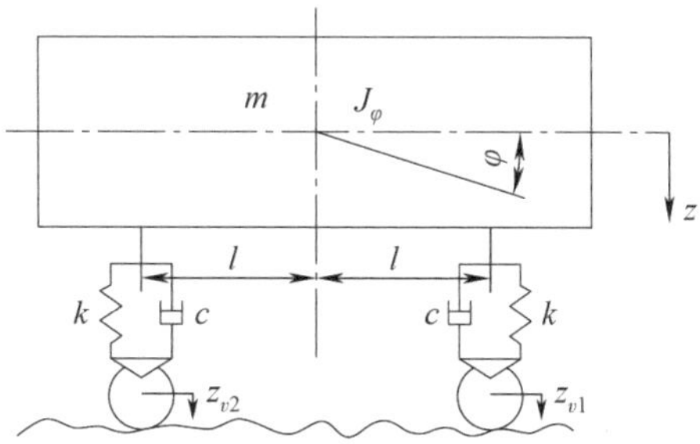

Fig. 7.2 Vertical vibration

When the front and rear parts of the vehicle body are symmetrical to the center of gravity, the float and nod vibrations are independent and uncoupled. Both vibrations can be characterized by the equations of motion of a SDOF system.

1. **Frequency Response Function of Vehicle Body Center of Gravity**

Considering the frequency response function of the vehicle body under the separate actions of z_f and z_s to be H_{z_f} and H_{z_s}, respectively, for the symmetric system shown in Fig. 7.2, the definition of the frequency response function is clearly $H_{z1}(\omega) = H_{z2}(\omega)$. By substituting $z_{v1} = e^{i\omega t}$ and $z_{v2} = 0$ into Eq. (7.4), we obtain

$$H_{z1}(\omega) = \frac{ic\omega + k}{-m\omega^2 + 2ic\omega + 2k} \tag{7.6}$$

The function can be expressed as a function of frequency and relative damping, and by substituting $P_1^2 = 2k/\mathrm{m}$ and $D_1 = c/(mP_1)$ into the Eq. (7.6), we get

$$H_{z1}(\omega) = H_{z2}(\omega) = \frac{1}{2}\frac{2iD_1P_1\omega + P_1^2}{P_1^2 - \omega^2 + 2iD_1P_1\omega} \tag{7.7}$$

According to the superposition principle of linear systems, when the system is subjected to multiple inputs, the response at a certain coordinate is the sum of the responses caused by each individual input. If we focus on $z_{v1} = e^{i\omega t}$ today, then $z_{v2} = e^{i\omega(t-\tau)}$ can be neglected. If the system response caused by z_{v1} and z_{v2} is $z_1(t)$ and $z_2(t)$, respectively, then the response of the body float and sink is

$$\begin{aligned} z(t) &= z_1(t) + z_2(t) = H_{z1}(\omega)e^{i\omega t} + H_{z2}(\omega)e^{i\omega(t-\tau)} \\ &= H_{z1}(\omega)(1 + e^{-i\omega t})e^{i\omega t} \\ &= H_{z1}(\omega)H_{z,in}(\omega)e^{i\omega t} = H_z(\omega)e^{i\omega t} \end{aligned} \tag{7.8}$$

where, $H_{z,in} = 1+e^{-i\omega t}$ is the input spacing function of floating and sinking vibration. The frequency response function of floating and sinking vibration of the vehicle body is $H_z(\omega) = H_{z1}(\omega)H_{z,in}(\omega)$.

Equation (7.8) indicates that the system can be treated as a single-input system with only one excitation if there is a delay τ between the two inputs, but the frequency response function $H_z(\omega)$ of the system is the vector sum of $H_{z1}(\omega)$ and $H_{z2}(\omega)e^{-i\omega\tau}$.

When the track is uneven, it is described by Formula (7.9), and $S_v(\Omega)$ can be converted to $S_v(\omega)$ as described above. Therefore, the response spectral density of the floating and sinking vibration of the two-axle railway vehicle is:

$$S_z(\omega) = |H_z(\omega)|^2 S_v(\omega) = |H_{z1}(\omega)|^2 \left|H_{z,in}(\omega)\right|^2 S_v(\omega) \tag{7.9}$$

and

$$\left|H_{z,in}(\omega)\right|^2 = \left|1 + e^{-i\omega\tau}\right|^2 = |1 + \cos(\omega\tau) - i\sin(\omega\tau)|^2 = 2[1 + \cos(\omega\tau)] \tag{7.10}$$

Similarly, the frequency response function $H_{\varphi1}(\omega)$ nodding of the nodding vibration of a two-axle railway vehicle can be obtained, so that $P_2^2 = 2kl^2/J_\varphi, \quad D_2 = cl^2/\left(J_\varphi P_2\right)$ can be obtained.

$$H_{\varphi1}(\omega) = -H_{\varphi2}(\omega) = \frac{1}{2l}\frac{2iD_2P_2\omega + P_2^2}{P_2^2 - \omega^2 + 2iD_2P_2\omega} \tag{7.11}$$

$$H_\varphi(\omega) = H_{\varphi1}(\omega)H_{\varphi,in}(\omega)$$

where, $H_{\varphi,in}(\omega)$ is the input spacing function of nodding vibration, and its expression is $H_{\varphi,\mathrm{in}}(\omega) = 1 - e^{i\omega\tau}$.

Thus, the response spectral density of the nodding vibration is

$$S_\varphi(\omega) = \left|H_\varphi(\omega)\right|^2 S_v(\omega) = \left|H_{\varphi1}(\omega)\right|^2 \left|H_{\varphi,in}(\omega)\right|^2 S_v(\omega) \tag{7.12}$$

$$\left|H_{\varphi,in}(\omega)\right|^2 = \left|1 - e^{-i\omega\tau}\right|^2 = 2[1 - \cos(\omega\tau)] \tag{7.13}$$

As observed from the above formulae, the input spacing function is related to frequency ω, velocity v, and τ, and its value can vary between 0 and 2. When $\omega\tau = \pi, 3\pi, 5\pi, \ldots$, the response of the body vibration corresponding to the ω is theoretically zero, while the $\left|H_{z,in}(\omega)\right|^2 = 0$ nodding of the body nodding vibration reaches its maximum. Conversely, when $\omega\tau = 0, 2\pi, 4\pi, \ldots$, the response of the car body nodding vibration is 0, while the response of floating and sinking vibration is at its maximum. Figure 7.3 illustrates the curve of input spacing function $\left|H_{\varphi,in}(\omega)\right|$ with $\omega\tau$. It should be noted that the above two spacing functions are only applicable to the response of the vehicle body's center of gravity and are not suitable for other positions.

The track input spectrum $S_v(\omega)$ of the vehicle is a continuous broadband spectrum, and there are always different ω values at the peak and valley values in Fig. 7.3

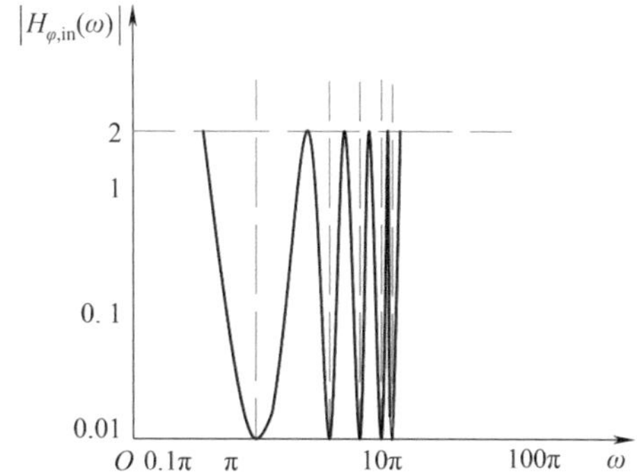

Fig. 7.3 Change of the spacing

at different running speeds. Consequently, the spectral density map of the vehicle response will exhibit fluctuations at these locations.

Given that the two-axle input functions of the vehicle float, sink, and nod vibration have a definite time-delay relationship, the calculation can be simplified by combining the two inputs with those on the front wheel. This method remains valid even when the vehicle is excited by more wheelsets.

2. **Vertical Random Response of a Railway Vehicle with Single Suspention**

(1) Motion Equation and Frequency Response Function

The model depicted in Fig. 2.25 represents a vehicle's vertical vibration system with a central suspension system. The four wheelsets are concurrently subjected to the input of track irregularities, and the vertical vibration of the vehicle body can be described by independent floating and sinking vibration equations as follows:

$$\boldsymbol{M}\ddot{\boldsymbol{Z}} + 2\boldsymbol{C}\dot{\boldsymbol{Z}} + 2\boldsymbol{K}\boldsymbol{Z} = \frac{c}{2}(\dot{z}_{v1} + \dot{z}_{v2} + \dot{z}_{v3} + \dot{z}_{v4}) + \frac{k}{2}(z_{v1} + z_{v2} + z_{v3} + z_{v4}) \quad (7.14)$$

$$\boldsymbol{J}_{\varphi}\ddot{\varphi} + 2\boldsymbol{C}l^2\dot{\varphi} + 2\boldsymbol{K}l^2\varphi = \frac{cl}{2}(\dot{z}_{v1} + \dot{z}_{v2} - \dot{z}_{v3} - \dot{z}_{v4}) + \frac{kl}{2}(z_{v1} + z_{v2} - z_{v3} - z_{v4}) \quad (7.15)$$

where, $z_{v1}, z_{v2}, z_{v3}, z_{v4}$ represent the vertical displacements of the first, second, third, and fourth wheelset relative to the track irregularity, respectively; $z_{v2} = z_{v1}(t - \tau_1)$, $z_{v3} = z_{v1}(t - \tau_2)$, $z_{v4} = z_{v1}(t - \tau_3)$; τ_1, τ_2, τ_3 are the delays of the second, third, and fourth wheelset relative to the first wheelset, with values given

$$\tau_1 = \frac{2l_1}{v}, \quad \tau_2 = \frac{2l}{v}, \quad \tau_3 = \frac{2(l_1 + l)}{v} \quad (7.16)$$

By setting $z_{v1} = e^{i\omega t}$ and the other three inputs to zero and substituting into Eq. (7.14), we obtain the frequency response function for the float and sink vibration due to a single input:

$$\boldsymbol{H}_{z1}(\omega) = \frac{1}{2}\frac{\mathrm{i}\boldsymbol{C}\omega + \boldsymbol{K}}{-\boldsymbol{M}\omega^2 + 2\mathrm{i}\boldsymbol{C}\omega + 2\boldsymbol{K}} \tag{7.17}$$

Comparing this with Eq. (7.7), it is evident that the frequency response function for a four-axle vehicle is only half that of a two-axle vehicle. This reduction is attributed to the fact that the displacement of the lower support point of the central spring is half the vertical input displacement of the wheelset after the implementation of a two-axle bogie. This indicates that employing a bogie configuration can mitigate the dynamic impact of track irregularities on the sprung mass.

Given the input z_{v1} of the first wheelset, the inputs for the remaining three wheelsets are:

$$\begin{cases} z_{v2} = \mathrm{e}^{\mathrm{i}\omega(t-\tau_1)} = z_{v1}\mathrm{e}^{-\mathrm{i}\omega\tau_1} \\ z_{v3} = \mathrm{e}^{\mathrm{i}\omega(t-\tau_2)} = z_{v1}\mathrm{e}^{-\mathrm{i}\omega\tau_2} \\ z_{v4} = \mathrm{e}^{\mathrm{i}\omega(t-\tau_3)} = z_{v1}\mathrm{e}^{-\mathrm{i}\omega\tau_3} \end{cases} \tag{7.18}$$

The response functions corresponding to these three distinct inputs are $\boldsymbol{H}_{z2}(\omega), \boldsymbol{H}_{z3}(\omega)$ and $\boldsymbol{H}_{z4}(\omega)$, respectively, due to the symmetry in the vehicle's structure and parameters. It follows that $\boldsymbol{H}_{z1}(\omega) = \boldsymbol{H}_{z2}(\omega) = \boldsymbol{H}_{z3}(\omega) = \boldsymbol{H}_{z4}(\omega)$.

The total response of the floating and sinking vibration for the four-axle vehicle body is the sum of the responses generated by each of the four inputs acting independently, which can be expressed as

$$\begin{aligned} z(t) &= \boldsymbol{H}_{z1}(\omega)\left(1 + \mathrm{e}^{-\mathrm{i}\omega\tau_1} + \mathrm{e}^{-\mathrm{i}\omega\tau_2} + \mathrm{e}^{-\mathrm{i}\omega\tau_3}\right)\mathrm{e}^{\mathrm{i}\omega t} \\ &= \boldsymbol{H}_{z1}(\omega)\boldsymbol{H}_{z,in}(\omega)\mathrm{e}^{\mathrm{i}\omega t} \\ &= \boldsymbol{H}_{z}(\omega)\mathrm{e}^{\mathrm{i}\omega t} \end{aligned} \tag{7.19}$$

Additionally,

$$\boldsymbol{H}_{z,in}(\omega) = 1 + \mathrm{e}^{-\mathrm{i}\omega\tau_1} + \mathrm{e}^{-\mathrm{i}\omega\tau_2} + \mathrm{e}^{-\mathrm{i}\omega\tau_3} = (1 + \mathrm{e}^{-\mathrm{i}\omega\tau_1})(1 + \mathrm{e}^{-\mathrm{i}\omega\tau_2}) \tag{7.20}$$

$$\boldsymbol{H}_{z}(\omega) = \boldsymbol{H}_{z1}(\omega)\boldsymbol{H}_{z,in}(\omega) \tag{7.21}$$

The above equations demonstrate that the four-axle vertical input with delay can still be treated as a single-input system with a single excitation, and the input spacing function must be considered within the system's frequency response function.

(2) Spectral Density and Mean Square Value of the Response

The spectral density of the floating and sinking vibration displacement for a four-axle railway vehicle is given by:

$$
\begin{aligned}
S_z(\omega) &= |\boldsymbol{H}_z(\omega)|^2 S_v(\omega) = |\boldsymbol{H}_{z1}(\omega)|^2 \left|\boldsymbol{H}_{z,in}(\omega)\right|^2 S_v(\omega) \\
&= \left|\frac{1}{2}\frac{\mathrm{i}\boldsymbol{C}\omega+\boldsymbol{K}}{-\boldsymbol{M}\omega^2+2\mathrm{i}\boldsymbol{C}\omega+2\boldsymbol{K}}\right|^2 \left|(1+\mathrm{e}^{-\mathrm{i}\omega\tau_1})(1+\mathrm{e}^{-\mathrm{i}\omega\tau_2})\right|^2 S_v(\omega) \\
&= [1+\cos(\omega\tau_1)][1+\cos(\omega\tau_2)]\left|\frac{\mathrm{i}\boldsymbol{C}\omega+\boldsymbol{K}}{-\boldsymbol{M}\omega^2+2\mathrm{i}\boldsymbol{C}\omega+2\boldsymbol{K}}\right|^2 S_v(\omega)
\end{aligned}
\tag{7.22}
$$

The spectral density $S_z(\omega)$ of the response is composed of the product of three transfer function ω terms. Since the form of the spectral density derived from actual track irregularities is not as straightforward as the previous Formula (7.9), it is generally impractical or impossible to solve for the vehicle response spectral density analytically. Instead, a numerical solution is typically sought using a computer to solve Eq. (7.22), following these steps:

① Determine the effective frequency range based on the spectral density $S_v(\omega)$ of the track irregularity and its cutoff frequency.
② Select a frequency ω_i within the effective bandwidth and determine the corresponding transfer function values.
③ Calculate the necessary coefficients and spectral densities for the given $|\boldsymbol{H}_{z,in}(\omega)|_i^2$.
④ Obtain the response spectral density $S_z(\omega)_i$ by multiplying the three components according to the formula (7.22).
⑤ Repeat steps ② and ③ with a new frequency ω_i selected from the effective bandwidth.
⑥ Accumulate the values to obtain the spectral density over the effective bandwidth.
⑦ The mean square values of the response are calculated as follows:

Displacement mean square value

$$
E(z^2) = \int_{\omega_{\min}}^{\omega_{\max}} S_z(\omega)\mathrm{d}\omega \tag{7.23}
$$

Mean square of acceleration

$$
E(\ddot{z}^2) = \int_{\omega_{\min}}^{\omega_{\max}} \omega^4 S_z(\omega)\mathrm{d}\omega \tag{7.24}
$$

For pitch vibration, the methods and steps for solving the frequency response function and spectral density are identical to those described above and are thus not detailed here. The response at the vehicle body's center plate is the vector sum of the responses from both float and nod vibrations, with the RMS values of displacement and acceleration given by:

$$
Z_{\sum \mathrm{RMS}} = \sqrt{Z_{\mathrm{RMS}}^2 + (l\varphi)_{\mathrm{RMS}}^2} \ddot{Z}_{\sum \mathrm{RMS}} = \sqrt{\ddot{Z}_{\mathrm{RMS}}^2 + (l\ddot{\varphi})_{\mathrm{RMS}}^2} \tag{7.25}
$$

3. **Vertical Random Response of Secondary Suspended Railway Vehicles**

The vertical vibration model of four-axle passenger vehicle is depicted in Fig. 2.25. To enhance vibration damping and absorption performance, two-axle bogies equipped with secondary suspension devices are widely utilized in railway vehicles as running units. The methodology for calculating the vertical response of four-axle vehicles under random irregularity excitation is analogous to the previous methods, with the system's DOF increased to six. The vehicle body and the two bogie frames each possess two DOF for floating and sinking as well as nodding vibration. The floating and sinking vibration of the vehicle body is coupled with that of the two frames, which can be described by a set of four DOF equations. The nodding vibrations of the two frames are independent and are characterized by two DOF equations.

The vibration equation of the railway vehicle body frame is expressed in matrix form as:

$$\boldsymbol{M}\ddot{z} + \boldsymbol{C}\dot{z} + \boldsymbol{K}z = \boldsymbol{C}_f\dot{z}_v + \boldsymbol{K}_j z_v \tag{7.26}$$

7.2 The Solution of Frequency Response Function

Railway vehicles are multi-input and multi-output systems. When the system has m inputs and n outputs, there are m frequency response functions corresponding to the output of each coordinate, and the entire system has a total of $m \times n$ frequency response functions $H_{ik}(\omega)(i = 1, 2, \cdots n, \quad k = 1, 2, \cdots m)$, excluding the frame's separate nod. For the model discussed here, there is $m = n = 4$, meaning the system has a total of 16 frequency response functions. Solving the vehicle response requires solving all $H_{ik}(\omega)$, which is more complex than the previous models, but the approach to solving the problem remains the same. Although there are numerous $H_{ik}(\omega)$, from an engineering calculation perspective, it is usually unnecessary to produce an analytical formula for each; instead, computer numerical solutions are employed to perform the numerical calculation of the response spectrum.

Figure 7.4 displays the spectral density of a vehicle body in China when operating on high-speed railway line using the track spectrum. The peak value in Fig. 7.3 appears at 0.9 Hz, corresponding to the low-frequency floating and sinking vibration of the vehicle. It is important to note that the curve fluctuation in Fig. 7.4 demonstrates the strong influence of the input spacing function on the spectral density curve, with the excitation peaks contained in the actual irregularity spectrum being amplified or reduced by the spacing function. In the 1–20 Hz range, the spectral density curve exhibits numerous peaks and valleys, determined by the curve shape of the input spacing function. Given the large axle box spring stiffness and light frame weight of the vehicle, the frame's resonance frequency is 30 Hz, but the corresponding peak in Fig. 7.5 is not prominent. This is because the input spacing function at 30 Hz is trough close to 0, which weakens the formant wave.

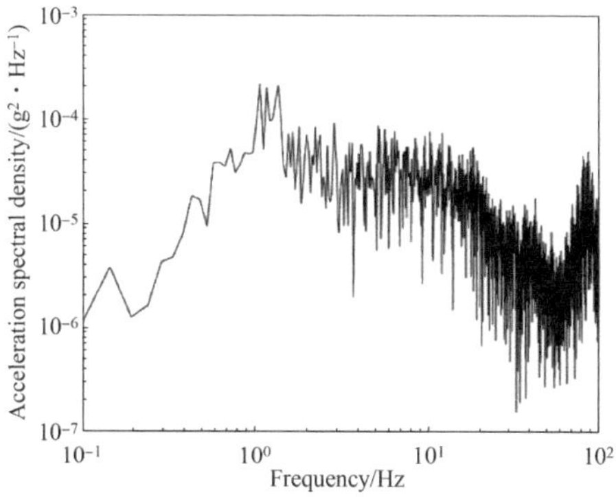

Fig. 7.4 Spectral density of vehicle body

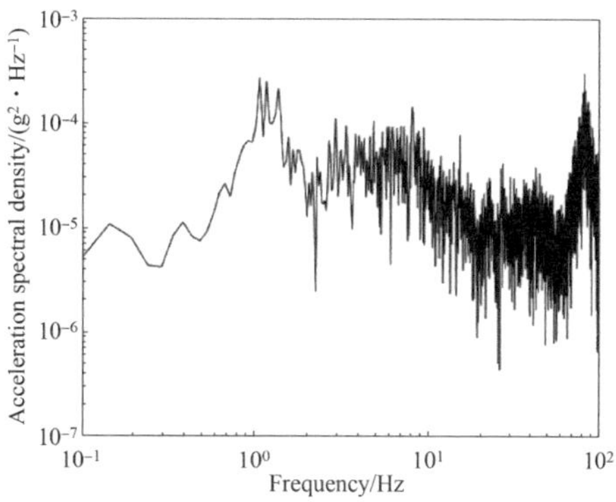

Fig. 7.5 Spectral density of resultant vertical vibration acceleration at the rear core plate of the vehicle body

Figure 7.5 presents the acceleration spectrum calculation diagram synthesized by float, sink, and nod vibration at the rear core plate of the vehicle body at the aforementioned speed. The diagram's line no longer exhibits the continuous peak-valley phenomenon as shown in Fig. 7.4. This is because the peak-valley values of the two input spacing functions of nod and float and sink vibration are staggered and compensate for each other. The RMS value of the resultant acceleration of the spectrum in Fig. 7.5 is 0.073 6 g. It is worth mentioning that the spectral density of orbital irregularity used in the response calculation in this example is the American spectrum with a coefficient $k = 1$. When $k = 1$ is taken, the theoretical excitation spectrum curve is biased towards the upper limit of the experimental value, leading to an overestimation of the calculated response acceleration value. When $k = 0.25$ is used for calculation, the RMS value of the resultant acceleration is halved, aligning well with the measured value.

7.3 Lateral Random Vibration of Railway Vehicles

Lateral random vibration of railway vehicles is caused by the random excitation of the track during operation, exhibiting characteristics of forced vibration. When the amplitude of the track direction irregularity is significantly greater than the wheel-rail clearance, intense vibration from the wheel rim hitting the rail will occur, distinct from the wheel-set instability snake motion described below. When operating below the critical speed of snaking instability, the snake motion of the railway vehicle on an ideal straight track gradually attenuates.

The transverse vibration of the wheelset caused by the track is transmitted to the vehicle body through the bogie, causing the bogie and the body to undergo continuous yaw, roll, and shake vibrations during operation. The lateral random vibration of vehicles will be discussed below.

7.3.1 *Establishment of Transverse Vibration Model of Railway Vehicle*

The primary difference between lateral and vertical vibrations of a vehicle lies in the wheel- rail interaction. In the previously mentioned vertical vibration model, there is no relative vertical movement between the wheel and the rigid track. In contrast, the lateral vibration model involves relative lateral movement between the wheelset and the rail.

The vibration equation of the vehicle under lateral random excitation of the track is

$$\boldsymbol{M}\ddot{y} + (\boldsymbol{C}_s + \boldsymbol{C}_r)\dot{y} + (\boldsymbol{K}_s + \boldsymbol{K}_r)y = \boldsymbol{B}\boldsymbol{u} + \boldsymbol{Q}\dot{\boldsymbol{u}} \tag{7.27}$$

where, $\boldsymbol{C}_{\mathrm{s}}$, $\boldsymbol{C}_{\mathrm{r}}$ and $\boldsymbol{K}_s$, K_{r} are the damping matrix and stiffness matrix resulting from the suspension system and wheel-rail interaction, respectively; $\boldsymbol{B}$ is the track input distribution matrix; $\boldsymbol{Q}$ is the distribution matrix of the input of the orbital horizontal irregularity derivative; $\boldsymbol{u}$ is the input vector for the orbital direction and horizontal irregularity. The other symbols retain their previous meanings.

If the right end of the Eq. (7.27) is $\mathbf{0}$, during operation in an ideal flat straight, it can be used to study the snake movement.

Equation (7.27) can predict various lateral dynamic responses of vehicles operating on real tracks, such as wheelset, bogie frame, and body displacement and vibration acceleration, suspension element travel, wheel-rail force, and response vector spectral density. The analysis indicates that calculating the lateral random response is more complex than the vertical random response. This complexity arises because the floating and sinking motion of the elastic track can be neglected in the vertical equation, with the wheelset moving along the vertical trajectory of the rigid track. The dynamic exciting force of the wheelset to the vehicle is generated by the deformation of a series of springs.

The component term of the lateral equation not only includes the lateral wheel-rail force caused by the relative movement between the wheel-rail, but also the wheelset's stimulation by the uneven direction and level of the track.

The following analysis will focus on constructing the incentive on the right side of the equal sign in Eq. (7.27).

The C116 Committee of the International Union of Railways (UIC) has comprehensively considered the influence of many factors when studying the horizontal excitation caused by track irregularity. They have derived the corresponding incentive force analysis formula and ultimately simplified it by omitting minor factors to obtain a more concise formula. In the discussion below, the effect of minor factors such as rotational creep is initially omitted to avoid tedious derivation, with both approaches yielding the same results.

Figure 7.6 illustrates the position of the track and wheel. The unconstrained free wheelset case is discussed first. Figure 7.6a indicates that when the track has no directional unevenness, the left and right wheels are in a central symmetric position when they contact the rail with the same radius. Figure 7.6b shows that when the track has directional unevenness, its centerline shifts from the ideal position to one side while the wheelset moves horizontally, and the rolling radii of the left and right wheels are r_l and r_r, respectively. Due to the tread's slope, the wheelset is accompanied by an angle θ^* when it moves across the track, and this value is constrained by the wheel-rail contact geometry.

$$\theta^* = -\frac{\sigma}{b}(y_w - y_n) \tag{7.28}$$

where, σ is called the roll coefficient, and for a cone tread, σ is equal to the slope of the tread; the wheelset angle θ^* will exert a force on the upper part of the running part through a series of vertical suspensions. Due to the small angle value, other effects can be ignored.

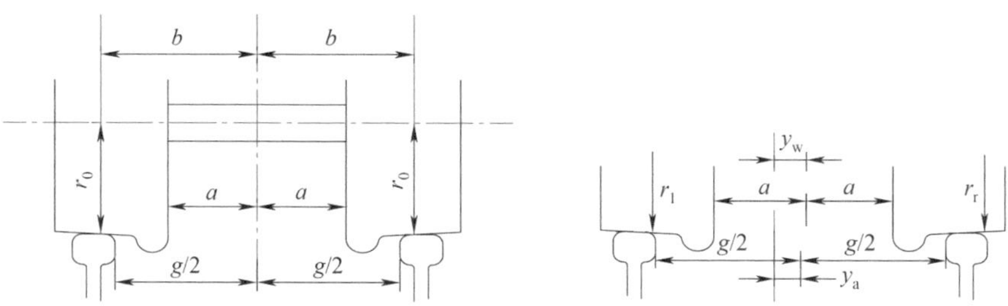

(a) No directional unevenness

(b) With directional unevenness

Fig. 7.6 Position of the track and wheel

When the track has horizontal irregularity, it also affects the suspension. The center of gravity of the wheelset will shift by $r_0\theta_c$, so the resultant rotation angle of the wheelset is

$$\theta_1^* = -\frac{\sigma}{b}(y_w - y_a - r_0\theta_c) \tag{7.29}$$

where, θ_1^* is the constrained angle of the connection between the relative two sides of the rail, and the metric basis of θ_c differs in the fixed coordinate system.

When the wheelset runs along the track with direction and horizontal irregularity and there is a shaking angle φ_w and transverse movement, the position of the wheel-rail contact point relative to the track section is constant. Then, the transverse velocity of the wheel-rail contact point relative to the material on the wheel is

$$\begin{aligned} v_{21} = v_{2r} = v_2 &= -(\dot{y}_w - r_0\dot{\theta}_1^*) + \dot{y}_a + v\varphi_w \\ &= (\dot{y}_a - \dot{y}_w)\left(1 + \frac{\sigma r_0}{b}\right) + \frac{\sigma r_0^2}{b}\theta_c + v\varphi_w \end{aligned} \tag{7.30}$$

At the same time, the transverse velocity of the contact spot on the track, that is, the transverse velocity of the wheel-rail contact point relative to the rail surface material,

$$v_{2t} = \dot{y}_a \tag{7.31}$$

Here, since the track is considered rigid and does not move laterally, although the direction and level are not even, it remains stationary while the wheelset rolls on it, producing relative geometric changes that affect the vertical and vertical creep force. The transverse creep speed at the wheel-rail contact spot can be considered as arising from the lateral speed of the wheelset. The derivative of the orbital direction irregularity does not produce a relative creep significance. After ignoring some minor factors, the lateral creep rate is defined as

$$v_2 = \frac{v_{2t} - v_2}{v} = \frac{1}{v}\left[\dot{y}_w\left(1 + \frac{\sigma r_0}{b}\right) - \frac{\sigma r_0}{b}\dot{y}_a - \frac{\sigma r_0^2}{b}\dot{\theta}_c - v\varphi_w\right] \tag{7.32}$$

It should be noted that it cannot be mistaken for the transverse relative speed difference between the wheel and rail. If the minor effect of rotational creep is excluded, the lateral creep force acting on each wheel contact spot is

$$T_y = -f_{22}v_{22} = -f_{22}\left(\frac{\dot{y}_w}{v} - \phi_w\right) \tag{7.33}$$

The force of the wheelset on the frame through vertical suspension can be expressed as $\pm\theta_1^* k_{1z} b$, and the rolling moment is $2k_{1z}b\theta_1^*$.

Due to the longitudinal creep caused by the track direction and horizontal irregularity, the longitudinal creep force of the left and right wheels is

$$\begin{cases} v_{1l} = \frac{\lambda_e}{r_0}(y_w - y_a - r_0\theta_c) + \frac{b\dot{\phi}_w}{v} \\ v_{1r} = \frac{-\lambda_e}{r_0}(y_w - y_a - r_0\theta_c) - \frac{b\dot{\phi}_w}{v} \end{cases} \tag{7.34}$$

where, λ_e is the equivalent wheel and rail conicity. Thus, the shaking head creep force moment acting on a wheelset is

$$M_z = -f_{11}v_{1l}b + f_{11}v_{1r}b = -2f_{11}\left[\frac{\lambda_e b}{r_0}(y_w - y_a - r_0\theta_c) + \frac{b^2\dot{\phi}_w}{v}\right] \tag{7.35}$$

Taking the aforementioned influence of gravity stiffness into account, the resilience, and the restoring moment acting on the wheelset are

$$\begin{cases} M_w\ddot{y}_w + 2f_{22}\left[\frac{\dot{y}_w}{v}\left(1 + \frac{r\sigma_0}{b}\right) - \phi_w\right] + K_{gy}y_w = 2f_{22}\left(\frac{\sigma r_0}{vb}\dot{y}_a + \frac{\sigma r_0^2}{vb}\dot{\theta}_c\right) + K_{gy}(y_a + r_0\theta_c) \\ J_w\ddot{\phi}_w + 2f_{11}\left(\frac{\lambda_e b}{r_0}y_w + \frac{b^2}{v}\dot{\phi}_w\right) - K_{g\phi}\phi_w = 2f_{11}\frac{\lambda_e b}{r_0}(y_a + r_0\theta_c) \end{cases} \tag{7.36a}$$

The actual calculation of the vehicle shows that the value of the σ item in Eq. (7.36a) is very small, with $\sigma r_0/b$ being approximately 0.03, so it can be omitted. Since the excitation effect of horizontal irregularity is significantly smaller than that of directional irregularity, it is often ignored in some practical calculations, leading to the simplified Eq. (7.36b).

$$\begin{cases} M_w\ddot{y}_w + 2f_{22}\left(\frac{\dot{y}_w}{v} - \phi_w\right) + K_{gy}y_w = K_{gy}y_a \\ J_w\ddot{\phi}_w + 2f_{11}\left(\frac{\lambda_e b}{r_0}y_w + \frac{b^2}{v}\dot{\phi}_w\right) - K_{g\phi}\phi_w = 2f_{11}\frac{\lambda_e b}{r_0}y_a \end{cases} \tag{7.36b}$$

If the equivalent slope is small and the effect of gravity stiffness can be neglected, Eq. (7.36b) becomes even simpler. However, since the wheelset is connected to the frame or side frame, when there is an elastic constraint at the connection, the effect of the elastic constraint should be considered, and the elastic force F_{s1} and the moment term M_{s1} should be added to the left of the equal sign of the above equation. Thus, Eq. (7.36b) becomes

$$\begin{cases} M_w\ddot{y}_w + F_{s1} + 2f_{22}\left(\frac{\dot{y}_w}{v} - \phi_w\right) + K_{gy}y_w = K_{gy}y_a \\ J_w\ddot{\phi}_w + M_{s1} + 2f_{11}\left(\frac{\lambda_e b}{r_0}y_w + \frac{b^2}{v}\dot{\phi}_w\right) - K_{g\phi}\phi_w = 2f_{11}\frac{\lambda_e b}{r_0}y_a \end{cases} \tag{7.36c}$$

where, F_{sl} and M_{s1} are determined by the vehicle model under study. For the truck without elastic suspension of the axle box, it is necessary to provide corresponding treatment at the connection between the axle box and the side frame according to the specific structure, and the constraint conditions there are force and torque added to the left of the equal sign of Eq. (7.36c).

When the connection between the wheelset and the frame is rigid positioning, the front and rear wheelsets and the frame (side) have no relative displacement in the transverse plane.

Therefore, the bogie as a whole will produce the yaw ϕ_t and the shaking head A motion around the core disc. Thus, under the incentive of the uneven track direction, the role of the above small factors is still ignored, and the transverse motion equation of the bogie is as follows.

$$\begin{cases} M_t\ddot{y}_t + F_{s2} + 4f_{22}\left(\frac{\dot{y}_t}{v} - \phi_t\right) + 2K_{gy}y_t = K_{gy}(y_{a1} + y_{a2}) \\ J_{t\phi}\ddot{\phi}_t + M_{s2} + \frac{4}{v}\left(f_{11}b^2 + f_{22}l_1^2\right)\dot{\phi}_t - 2K_{g\phi}\phi_t + 4f_{11}\frac{\lambda_e b}{r_0}y_t = 2f_{11}\frac{\lambda_e b}{r_0}(y_{a1} + y_{a2}) \end{cases} \tag{7.37}$$

where, M_t and $J_{t\phi}$ are the mass of the whole bogie and the shaking head moment of inertia around the core disc, respectively; F_{s2} and M_{s2} are the force and torque of the elastic suspension system acting on the bogie by the vehicle body; y_{a1} and y_{a2} indicate the track direction irregularity at the front and rear wheelsets, respectively. Other symbols retain their previous meanings.

Thus, the excitation matrix to the right of the equal sign of Eq. (7.27) can be determined by the excitation term of Eqs. (7.36c) or (7.37), and the entire system of Eq. (7.27) can be obtained.

7.3.2 *The Solution of Frequency Response Function and Spectral Density*

Since the form of Eq. (7.27) is the same as that of Eq. (7.14), the method and steps of solving the vertical frequency response function in Eq. (7.14) can be applied to solve the transverse vibration frequency response function, and then each output vector of the system can be obtained. For the above horizontal excitation, there is also the same deterministic delay relationship between the wheelsets as the vertical excitation, just as there is a definite time τ behind y_{a1} in Eq. (7.37). Therefore, the input received by each wheelset in the system is converted into the input processing with only one front guide wheelset using the input spacing function described above, and the method is the same as that described in the previous section.

If the excitation of horizontal irregularity also needs to be calculated, the frequency response function $H_{in}(\omega)$ and $H_\theta(\omega)$ corresponding to the direction and horizontal irregularity excitation should be found, respectively.

The power spectral density of any response vector in the system under the action of orbital lateral irregularity is:

$$S_w = \left|H_y(\omega)\right|^2 S_y(\omega) + H_y^*(\omega)H_\theta(\omega)S_{y\theta}(\omega)$$

$$+ H_\theta^*(\omega) H_y(\omega) S_{\theta y}(\omega) + |H_\theta(\omega)|^2 S_\theta(\omega) \tag{7.38}$$

where, $S_y(\omega)$ and $S_\theta(\omega)$ represent the input spectral density of the track direction and horizontal irregularity, respectively; $S_{\theta y}(\omega)$ is the input cross-spectral density.

Research indicates that the correlation between the direction and horizontal irregularity of the orbits is small, and their mutual spectral densities are small compared with their respective self-spectral densities, which can be omitted, so the Eq. (7.38) can be simplified as:

$$S_\mathrm{w} = \left|H_y(\omega)\right|^2 S_y(\omega) + |H_\theta(\omega)|^2 S_\theta(\omega) \tag{7.39}$$

In this way, the response spectral density of the two can be calculated separately, and the output mean square value of each response is the sum of the mean square values of the two.

7.4 Lateral Random Response Characteristics of Railway Vehicles

An overview of the common response characteristics of ordinary railway vehicles is provided. The primary suspension system of the bogie used in the vehicle is a metal rubber spring, the secondary suspension system is a circular spring and an air spring in parallel, and its lateral flexibility is determined by the stiffness of the circular spring and the lateral stabilizer bushing. The train operates at a speed of 176 km/h on the Level 6 line.

The wheelset produces forced lateral vibration under the action of track direction and level irregularity. Supported by inertia force and suspension force, the wheelset attempts to maintain its original motion state, and the geometric deviation of track position with the relative change of tread will cause snake motion. Consequently, the vehicle self-excited snake and forced response coexist. On the other hand, the lateral relative movement of wheel and rail will also cause the wheel rim to touch the rail, resulting in rigid impact, so it is necessary to check whether the offset exceeds the wheel and rail clearance. The lateral offset of the wheelset with respect to the rail is

$$y_\mathrm{e}(t) = y_\mathrm{w}(t) - y_\mathrm{a}(t) \tag{7.40}$$

To solve $y_\mathrm{e}(t)$, it is necessary to convert the spectral density of the input track direction irregularity from the frequency domain to the time domain function. This involves converting the spectral density into the function of time history $y_\mathrm{a}(t)$ or mileage and using this as an incentive function to solve the time history of wheelset relative transverse movement $y_\mathrm{e}(t)$ by numerical integration method.

When solving the frequency response function of the wheelset with respect to the transverse movement $y_\mathrm{e}(t)$, the Eq. (7.40) or its simplified formula should be replaced by $y_e(t)$ and $y_a(t)$. The frequency response function of the front guide wheelset under the excitation of track direction irregularity is shown in Fig. 7.7. It can be observed that when the movement frequency of the wheelset is lower than 1 Hz, the frequency response function is less than 0.2, indicating that the relative motion is small and the tracking capability of the wheelset is strong.

However, there is a peak region between 2 and 3 Hz, with a peak value at 2.8 Hz, where the relative movement of the wheelset is amplified. The analysis shows that it is an extended interval of snaking, and 2.8 Hz corresponds to the geometric snaking frequency of the wheelset ($v = 176\,\text{km/h}$). At higher frequency bands, the frequency response function of relative motion is 1, indicating that the wheelset is primarily in snaking motion and does not vibrate with the track irregularity. Figure 7.8 displays the spectral density curve of the front guide wheel against $y_e(t)$, and it can be seen that the mean square value of $y_e(t)$ is mainly clustered around 2. 8 Hz.

Figure 7.9 presents the waveforms of the offset motion for the front wheelset, where ($\lambda = 0.1$), at various speeds.

Figure 7.9a illustrates that when $k = 0.25$ and the operating speed is half of the critical speed, the RMS value of the wheelset offset is significantly reduced. This

Fig. 7.7 Frequency response of the front wheelset relative to the lateral offset

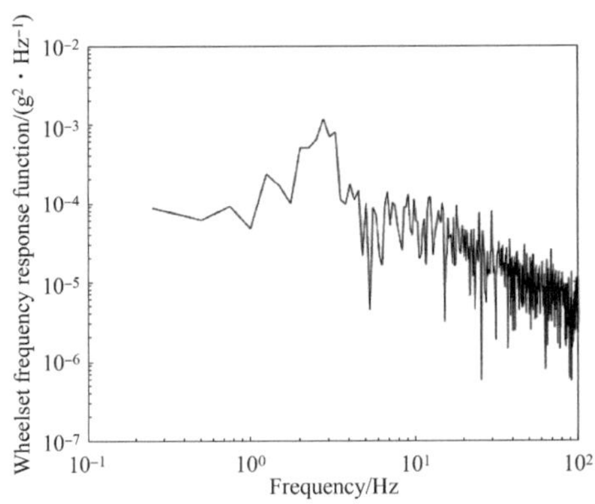

Fig. 7.8 Spectrum density of lateral migration of the front wheelset

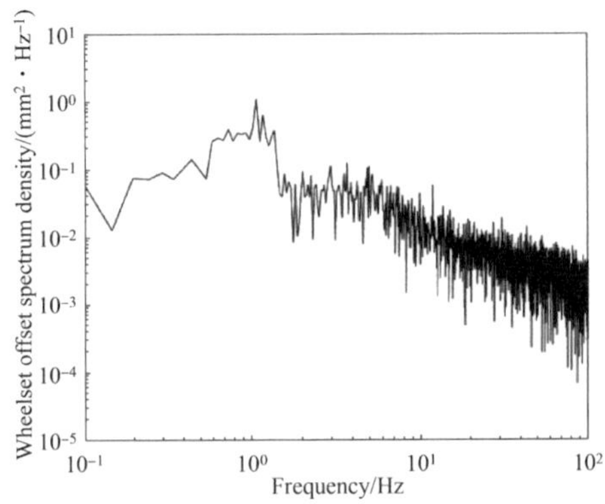

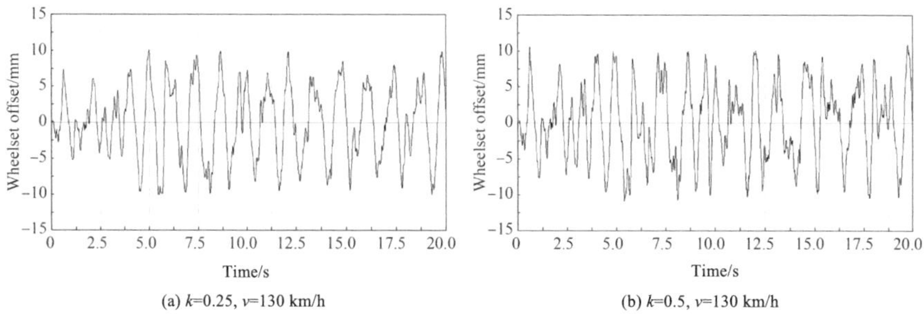

Fig. 7.9 Offset motion waveform of the front wheelset

indicates that better track conditions, which correspond to smaller amplitudes of track irregularities in the direction of travel, result in reduced wheelset offsets, and vice versa. Figure 7.9b demonstrates that when $k = 0.5$ and the operating speed surpasses the critical value, the wheelset offset exceeds the wheel-rail clearance, often leading to the phenomenon known as wheel flange climbing. At this juncture, the dynamic behavior of the wheelset has transitioned from forced vibration at lower speeds to a state where both forced vibration and snake instability coexist, with the motion waveform being the composite outcome of these two types of vibrations. Consequently, if the track is excessively irregular, it can still engender a vicious cycle of frequent flange impacts on the rail, even if the train's operating speed is below the critical speed.

Exercise 7

E7.1: What is the railway unevenness?

E7.2: What is the fundamental reason that wheel-rail system differ from classical mechanical system?

E7.3: Using wheels and rails as the interaction system, discuss the root cause of the internal instability tendency in locomotive and vehicle systems.

E7.4: In the hunting motion equation of the wheelset elastic positioning bogie, the wheelset has two degrees of freedom: lateral sway and yaw. What are the three forces that affect the motion of these two DOF?

E7.5: The mathematical expression used to evaluate the operational quality of the vehicle itself and the ride index of passenger comfort. What factors account for the difference between the two?

E7.6: Let the track spectral density be $S(f) = g/f^2$, and when the vehicle speed is v, convert it into a function $S(w)$ of time frequency w.

Bibliography

1. LALANNE C. Random vibration: Volume III [M]. London: Hermes Penton Ltd., 2002.
2. ZHU W Q. Random vibration[M]. Beijing: Science Press, 1992.
3. GAN Y C, XIE S H. Basic theory and applications of random vibration[M]. Changsha: Hunan Science and Technology Press, 1982.
4. HU Y S. Modern rail vehicle dynamics[M]. Beijing: China Railway Publishing House, 2009.

J. Zhang and H. Zhou, *Random Vibration and Spectrum Analysis*,
https://doi.org/10.1007/978-981-95-0191-5

MIX
Papier aus verantwortungsvollen Quellen
Paper from responsible sources
FSC® C105338
FSC
www.fsc.org

If you have any concerns about our products,
you can contact us on
ProductSafety@springernature.com

In case Publisher is established outside the EU,
the EU authorized representative is:
Springer Nature Customer Service Center GmbH
Europaplatz 3, 69115 Heidelberg, Germany

Printed by Libri Plureos GmbH
in Hamburg, Germany